Oliver Gywat, Hubert J. Krenner,
and Jesse Berezovsky
Spins in Optically Active
Quantum Dots

Related Titles

Schleich, W. P., Walther, H. (eds.)

Elements of Quantum Information

2007
ISBN 978-3-527-40725-5

Bruß, D., Leuchs, G. (eds.)

Lectures on Quantum Information

2007
ISBN 978-3-527-40527-5

Vogel, W., Welsch, D.-G.

Quantum Optics

2006
ISBN 978-3-527-40507-7

Stolze, J., Suter, D.

Quantum Computing
A Short Course from Theory to Experiment

2004
ISBN 978-3-527-40438-4

Bachor, H.-A., Ralph, T. C.

A Guide to Experiments in Quantum Optics

2004
ISBN 978-3-527-40393-6

Oliver Gywat, Hubert J. Krenner, and Jesse Berezovsky

Spins in Optically Active Quantum Dots

Concepts and Methods

WILEY-VCH

WILEY-VCH Verlag GmbH & Co. KGaA

The Authors

Dr. Oliver Gywat
University of California at Santa Barbara
California Nanosystems Institute
Santa Barbara, California 93106
USA
and
Credit Suisse Private Banking
Investment Services and Products
Bleicherweg 33
8070 Zürich
Switzerland

Dr. Hubert J. Krenner
Universität Augsburg
Institut für Physik
Lehrstuhl für Experimentalphysik I
Universitätsstr. 1
86159 Augsburg
Germany

Dr. Jesse Berezovsky
Harvard University
McKay Laboratory
9 Oxford St.
Cambridge, MA 02138
USA

Library of Congress Card No.: applied for

British Library Cataloguing-in-Publication Data:
A catalogue record for this book is available from the British Library.

Bibliographic information published by the Deutsche Nationalbibliothek
The Deutsche Nationalbibliothek lists this publication in the Deutsche Nationalbibliografie; detailed bibliographic data are available on the Internet at http://dnb.d-nb.de.

© 2010 WILEY-VCH Verlag GmbH & Co. KGaA, Weinheim

Printed in Great Britain
Printed on acid-free paper

Typesetting le-tex publishing services GmbH, Leipzig
Printing and Binding T.J. International Ltd., Padstow, Cornwall
Cover Design Adam Design, Weinheim

ISBN 978-3-527-40806-1

Contents

Spins in Optically Active Quantum Dots. Concepts and Methods.
Oliver Gywat, Hubert J. Krenner, and Jesse Berezovsky
Copyright © 2010 WILEY-VCH Verlag GmbH & Co. KGaA, Weinheim
ISBN: 978-3-527-40806-1

Preface

Nearly 50 years since the discovery of molecular beam epitaxy, extraordinary discoveries in science and technology have been fueled by the ability to fabricate matter with exquisite precision at the atomic level. This capability of producing "designer materials" has catalyzed fundamental scientific discoveries ranging from two-dimensional electron gases to the quantum and fractional quantum Hall effects, and has helped launch technological revolutions driven by the development of transistors and lasers. In addition to preparing carefully crafted thin films with engineered electronic characteristics, this class of material growth techniques coupled with the revolutionary concept of semiconductor heterostructures has enabled researchers to synthesize novel two-dimensional quantum wells and superlattices, and subsequently transfer their implementations into a myriad of electrical and optical devices. By reducing the dimensionality of three-dimensional matter and thus laterally confining electrons, a variety of new physical properties and carrier efficiencies have emerged in electronics. Many of these material systems have rapidly moved from the scientific arena to the technological world with significant impacts throughout society. From information processing to lighting, the rapid progress of semiconductor science and technology driven by artificially changing the dimension of matter raises an exciting question: what will the future bring as we continue to push towards the limit of zero dimensions?

This book addresses the new frontier of zero-dimensional matter and the manifest opportunities for science and engineering. While "no man is an island", individual electrons thrive in isolation when localized in semiconductor nanostructures. Their lifetimes are often longer and their optical emissions typically brighter than their peers in bulk materials. In a natural progression from thin films to quantum wells to quantum wires, the availability of semiconductor quantum dots offers a powerful laboratory within which to explore the limits of quantum confinement and the emergence of well-defined individual electronic states. Commensurate with this additional degree of confinement, smaller dimensions bring a transition from the classical to the quantum world, along with fundamental questions regarding the preparation, manipulation, and measurement of these states. These developments also raise the spectre of next-generation quantum mechanical electronics and its potential for affecting a discontinuous change in technology. While the electron charge has been the basis for today's devices, this new regime allows

Spins in Optically Active Quantum Dots. Concepts and Methods.
Oliver Gywat, Hubert J. Krenner, and Jesse Berezovsky
Copyright © 2010 WILEY-VCH Verlag GmbH & Co. KGaA, Weinheim
ISBN: 978-3-527-40806-1

us to consider the role of quantum variables such as the electron spin for additional control in future systems. These flexible structures serve as a common point for research at the intersection of quantum electronics, quantum optics, spintronics, and quantum computation that currently helps drive the field of quantum information science.

Beginning with an overall introduction that helps motivate the field, this book lays a scientific foundation by discussing the basic principles of spin physics and the broad range of quantum dot structures that have been grown using a variety of materials and epitaxial techniques. The authors subsequently delve into a discussion of several schemes for integrating quantum dots into electrical and optical devices, new phenomena that arise upon coupling quantum dot structures to one another, and then present salient experimental results, including specific measurement techniques. A substantial portion of the text is spent developing the theory of spins in quantum dots and spin-based optics, as well as exploring multifunctional systems that integrate electronics and photonics in a single device. Furthermore, the authors discuss the unusual role of nuclear spins in quantum dots and their potential for modifying spin-based phenomena.

If history provides any insight, the emergence of nanometer-scale quantum mechanical devices will be accompanied by a new generation of quantum scientists and engineers who will require a unique set of resources. This volume represents a thoughtfully prepared combination of experimental and theoretical information that will serve as an informative perspective for both present and future researchers in the field.

Santa Barbara, California, USA, April 2009 *David D. Awschalom*

1
Introduction

At the scale of nanometers, approaching the size of the fundamental constituents of materials, various subdisciplines of physics merge. To provide a closed description of many nanoscale phenomena, considerations from varied fields must be taken into account, such as quantum mechanics, optics, quantum optics, semiconductor physics, material science, atomic and molecular physics, and so on. Crucially, theory and experiments must go hand in hand in the discovery and elucidation of such effects. In this book, we present one example of such nanoscale science – the physics of spins in optically active quantum dots. The variety of areas of physics that must be brought to mind to understand these systems can be seen in the structure of this book, which contains chapters on materials growth and synthesis, solid state and quantum optics theory, and experimental methods.

This chapter will introduce the two key terms in the title: spin, the fundamental angular momentum of a particle, and nanometer-sized semiconductor structures, known as quantum dots. Specifically, we focus on quantum dots whose properties can be measured and controlled via their interaction with light, and how spins in these structures may be investigated and potentially used for novel applications such as quantum information processing. We expand the discussion on quantum dots in Chapter 2 by describing the two main fabrication techniques, semiconductor epitaxy and chemical synthesis. This is followed by some theoretical background on semiconductor physics and confined states in different types of quantum dots in Chapter 3. We then show in Chapter 4 that semiconductor diodes and optical cavities can be used to provide the knobs required to control the electrical, optical, and spin properties of optically active quantum dots for applications. To back up all the experimental findings and techniques, Chapter 5 provides the elementary theory of the coupling between confined states to electromagnetic radiation. The interactions between spins of carriers and a carrier and the nuclei in the dot's crystal lattice are then discussed in Chapter 6, before we switch back to experimental techniques. The rich toolkit to initialize, manipulate, and read out spins in quantum dots by optical means is opened and explained in Chapter 7. In the concluding Chapter 8 we will add another important part to this discussion, namely the coupling of quantum dots. In coupled quantum dots the interactions between charges and spins show a subtle interplay and provide us with the potential to use these optically active nanostructures for a scalable architecture for

Spins in Optically Active Quantum Dots. Concepts and Methods.
Oliver Gywat, Hubert J. Krenner, and Jesse Berezovsky
Copyright © 2010 WILEY-VCH Verlag GmbH & Co. KGaA, Weinheim
ISBN: 978-3-527-40806-1

quantum information processing and also to observe fundamental phenomena in coupled quantum dots, analogous to the effects of coupled atoms in molecular physics.

1.1
Spin

Spin, the intrinsic angular momentum of a particle, was first described theoretically by George Uhlenbeck and Samuel Goudsmit in 1925, and formalized by Wolfgang Pauli in 1926. Experimentally, however, spin phenomena have been observed and put to practical use for much longer. The earliest known spin-based device is most likely the magnetic compass. Here, a freely rotating needle is constructed out of a material in which electron spins align with each other under their mutual exchange interaction. This leads to a macroscopic spin polarization in the needle (ferromagnetism), causing the needle to align with the Earth's magnetic field due to the Zeeman interaction of a spin with a magnetic field (see Eq. (1.5)). Written records from ancient China referring to these phenomena date back to the fourth century BC.

Over the next 2300 years or so, the knowledge of magnetism spread across the globe. Clever minds devised new uses for the phenomenon, and refined old ones, ranging from the electric motor, to the dynamo, to the posting of notes on a refrigerator. Despite its bountiful technological applications, at the beginning of the twentieth century, the physical origins of magnetism were still unclear.

As quantum mechanics was being developed in the 1920s, great strides were made in understanding atomic spectra by quantizing the orbital momentum of electrons around the atomic nucleus. However, results such as the Stern–Gerlach experiment, and unexplained splittings in atomic spectra (the "anomalous Zeeman effect", and hyperfine splitting) indicated that there were extra quantum degrees of freedom not being taken into account.

A natural candidate for this unknown quantity was the angular momentum of a particle. The idea was at first considered to be impossible. Given the known upper bound on the radius of the electron, the angular velocity of the electron would need to be impossibly high to provide the observed splittings. Nevertheless, Uhlenbeck and Goudsmit published the idea in 1925. Despite its apparent impossibility, the idea of "spin" nicely explained the observations. Originally a skeptic, Wolfgang Pauli warmed to the idea and ran with it, redefining spin not as an actual rotation of a particle, but as an angular momentum intrinsic to the particle, just as charge or mass are intrinsic properties. He then went on to develop a formalism for dealing with spin in (nonrelativistic) quantum mechanics (see, e.g., [1]).

Once this theoretical framework was in place, the experimental study of spin physics could now proceed hand-in-hand with theory, instead of the pure phenomenology of the previous millennia. Throughout the rest of the twentieth century numerous advances were made, such as a detailed understanding of magnetic materials, nuclear spin physics, and spin resonance phenomena. These discoveries led to revolutionary technologies such as magnetic resonance imaging (MRI) and magnetic data storage (tapes, hard drives).

The reservoir of interesting spin phenomena is still far from dry. Recent advances in materials, electronics, and low temperature technologies have brought new untapped wells of spin physics within reach. One of the fruits of these new capabilities has been the development of quantum dots, which creates a straightforward way to isolate single or few spins for study or possible applications.

1.2
Spin-1/2 Basics

According to quantum mechanics, angular momentum as an observable can be described by two quantities: the angular momentum quantum number l and the projection of angular momentum on the (say) z axis, m. Throughout this book, we do not indicate operators with any special notation, assuming that the reader is familiar with the basics of quantum mechanics and that it is clear from the context which quantities are operators. The angular momentum quantum numbers are just the eigenvalues of the commuting operators $\mathbf{L}^2$ and L_z,

$$\mathbf{L}^2|\psi\rangle = l(l+1)\hbar^2|\psi\rangle \quad \text{and} \quad L_z|\psi\rangle = m\hbar|\psi\rangle , \tag{1.1}$$

where $\mathbf{L}^2 = L_x^2 + L_y^2 + L_z^2$, with L_α the angular momentum operator along the α direction, and $|\psi\rangle$ is a quantum mechanical state in Dirac notation. The quantum number l can take on half-integer values, and for a given l, the projection m can take on values $m = -l, -l+1, \ldots, l$.

In the case of a particle's spin, we consider an internal angular momentum with fixed quantum number s, and the projection of the spin can take on $2s+1$ values, from $-s$ to s. An electron has total spin $s = 1/2$ and projections $m_s = \pm 1/2$. Therefore, there are two eigenstates for $s = 1/2$, one with $m_s = +1/2$ denoted $|\uparrow\rangle$ and the other with $m_s = -1/2$ denoted $|\downarrow\rangle$. A general spin state of an electron is then given in a two-dimensional Hilbert space by a superposition of "spin up" and "spin down" states,

$$|\psi\rangle = \alpha|\uparrow\rangle + \beta|\downarrow\rangle , \tag{1.2}$$

where α and β are complex numbers satisfying the normalization condition $|\alpha|^2 + |\beta|^2 = 1$.

In the "spin up" and "spin down" basis, it is convenient to represent the operators S_α in matrix form,

$$S_x = \frac{\hbar}{2}\begin{pmatrix} 0 & 1 \\ 1 & 0 \end{pmatrix} \quad S_y = \frac{\hbar}{2}\begin{pmatrix} 0 & -i \\ i & 0 \end{pmatrix} \quad S_z = \frac{\hbar}{2}\begin{pmatrix} 1 & 0 \\ 0 & -1 \end{pmatrix}.$$

In this representation, the vector $(\alpha, \beta)^T$ represents the state given by Eq. (1.2), where T indicates transposition.

The above matrices without the factor of $\hbar/2$ are known as Pauli matrices, and the vector $\mathbf{S} = (S_x, S_y, S_z)$ is the spin operator for the electron spin-1/2.

Note that there are four degrees of freedom in the two complex coefficients α and β in Eq. (1.2). However, the normalization requirement removes one of these degrees of freedom, and another one is an overall phase that can be ignored as it cancels in matrix elements whenever an expectation value of an observable is calculated. Thus there are only two degrees of freedom that we care about, and Eq. (1.2) can be rewritten in the form

$$|\psi\rangle = \cos\frac{\theta}{2}|\uparrow\rangle + e^{i\phi}\sin\frac{\theta}{2}|\downarrow\rangle \ . \tag{1.3}$$

The two parameters θ and ϕ can be thought of as the polar and azimuthal angles defining a point on a sphere. This is known as the Bloch sphere, shown in Figure 1.1, and turns out to be a very useful way of picturing a spin 1/2 or any other two-level quantum system. The usefulness of this picture can be seen by looking at the expectation values of the spin in the x, y, and z directions. Using the matrix forms of the S_α operators given above, it is easy to show that the corresponding expectation values are

$$\langle S_x \rangle = \frac{\hbar}{2}\cos\phi\sin\theta \quad \langle S_y \rangle = \frac{\hbar}{2}\sin\phi\sin\theta \quad \langle S_z \rangle = \frac{\hbar}{2}\cos\theta \ . \tag{1.4}$$

These expectation values are equivalent to the x, y, and z components of the Bloch vector, as shown in Figure 1.1. Therefore, it is correct in some sense to think of the spin as actually "pointing" along the vector on the Bloch sphere. This one-to-one correspondence between the quantum state in a two-dimensional Hilbert space and the intuitive picture of a classical angular momentum vector in Euclidean space is apparently just a coincidence. Note, however, that the corresponding groups acting on the spin in these spaces, SU(2) and SO(3), are not isomorphic [1], giving room for purely quantum phase effects not captured in the Bloch

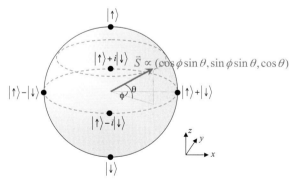

Fig. 1.1 The Bloch sphere. The vectors pointing to the north and south poles of the Bloch sphere represent the "up" and "down" eigenstates, with the rest of the sphere representing superpositions of "up" and "down".

sphere picture. For spins larger than 1/2 it is harder to think of a corresponding visualization in real space. But since here we are typically interested in electron spins, the Bloch sphere provides a useful and intuitive way of thinking about quantum spin states.

A spin in a magnetic field **B** has a contribution to its energy from the Zeeman Hamiltonian,

$$H_Z = \frac{g\mu_B}{\hbar}\mathbf{B}\cdot\mathbf{S}, \tag{1.5}$$

where $\mu_B = 9.274 \times 10^{24}$ J/T is the Bohr magneton. For an electron in vacuum, the electron gyromagnetic factor or g-factor is approximately $g = 2$. However, this is not a universal property. The spin–orbit interaction modifies this quantity, in some crystals even up to an extent that its sign is reversed. For example, electrons in the conduction band of GaAs have $g = -0.44$.

The result of the Zeeman effect on an electron spin is clearly seen by choosing the z axis to be along the magnetic field, leading to

$$H_Z = \frac{1}{2}g\mu_B B_z \begin{pmatrix} 1 & 0 \\ 0 & -1 \end{pmatrix}. \tag{1.6}$$

The spin eigenstates $|\uparrow\rangle$ and $|\downarrow\rangle$ are split by the Zeeman splitting $\Delta E = g\mu_B B_z$. If the spin is not in an eigenstate, then it evolves in time, depending on the Zeeman splitting. For a spin in the state given by Eq. (1.3) at $t = 0$, the state evolves according to (again, ignoring the overall phase)

$$|\psi(t)\rangle = \cos\frac{\theta}{2}|\uparrow\rangle + e^{i(\omega_L t + \phi)}\sin\frac{\theta}{2}|\downarrow\rangle, \tag{1.7}$$

where $\hbar\omega_L = g\mu_B B_z$ is the Zeeman splitting. The angular frequency ω_L is known as the Larmor frequency. In the Bloch sphere picture, this corresponds to the spin vector precessing about the z axis at the Larmor frequency,

$$\mathbf{S}(t) = (\cos(\omega_L t + \phi)\sin\theta, \sin(\omega_L t + \phi)\sin\theta, \cos\theta). \tag{1.8}$$

This phenomenon is referred to as Larmor precession.

If perfectly isolated from the environment, a spin in a static magnetic field would obey the dynamics described above forever. In reality, there are a number of effects that damp the evolution in time of an electron spin in a semiconductor. These effects can be divided into two categories: those that randomize the relative phase ϕ, and those that affect θ in Eq. (1.7). The randomization of θ is referred to as longitudinal spin relaxation, and is characterized by a time T_1. The loss of the relative phase information ϕ is referred to as transverse spin decoherence, occurring in time T_2. To illustrate these two time scales, we can say that a spin prepared in the excited state will relax into equilibrium on the time scale T_1. A spin that precesses in the plane of the equator of the Bloch sphere, as for $\theta = \pi/2$ in Eq. (1.7), will be distributed randomly on the equator after the characteristic time T_2.

This type of damping of the Larmor precession is taken into account in the Bloch equations,

$$\dot{\mathbf{S}}(t) = \mathbf{S} \times \mathbf{h} - R(\mathbf{S} - \mathbf{S}_\infty) \,. \tag{1.9}$$

Here, the first term on the right-hand side describes the precession of the spin components due to a magnetic field $\mathbf{B}$ along z, contained in $\mathbf{h} = (0, 0, \omega_L)$, where ω_L is again the Larmor frequency. The second term describes relaxation and decoherence of $\mathbf{S}$ towards the equilibrium spin polarization $\mathbf{S}_\infty = (0, 0, \tilde{S})$, which occurs due to

$$R = \begin{pmatrix} 1/T_2 & 0 & 0 \\ 0 & 1/T_2 & 0 \\ 0 & 0 & 1/T_1 \end{pmatrix} . \tag{1.10}$$

In this description it is intuitively clear that the decoherence time T_2 is called the transverse spin lifetime, as it acts on the transverse spin components, S_x and S_y. The relaxation time T_1, in turn, affects the z component and is therefore called the longitudinal spin lifetime. The single-spin Bloch equation can be written in the more compact form

$$\dot{\mathbf{S}}(t) = -\Omega \, (\mathbf{S} - \mathbf{S}_\infty) \,, \tag{1.11}$$

since $\mathbf{h} \times \mathbf{S}_\infty = 0$. The solution of Eq. (1.9) is now just given by

$$\mathbf{S}(t) = e^{-\Omega t}\mathbf{S}(0) + (1 - e^{-\Omega t})\mathbf{S}_\infty \,, \tag{1.12}$$

with the components

$$S_x(t) = S_x(0)e^{-t/T_2}\cos(\omega_L t) + S_y(0)e^{-t/T_2}\sin(\omega_L t) \,, \tag{1.13}$$

$$S_y(t) = -S_x(0)e^{-t/T_2}\sin(\omega_L t) + S_y(0)e^{-t/T_2}\cos(\omega_L t) \,, \tag{1.14}$$

$$S_z(t) = S_z(0)e^{-t/T_1} + \tilde{S}\left(1 - e^{-t/T_1}\right). \tag{1.15}$$

In the above solution of the Bloch equations we can nicely see the decay of the precession amplitude with the characteristic time T_2 and the relaxation into an equilibrium spin polarization $\tilde{S}$ along z with the characteristic time T_1.

As a caveat we would like to mention that not all baths that damp the evolution of a spin lead to nice exponential decays, as implicitly assumed by the above characteristic decay times T_1 and T_2. For example, the bath of nuclear spins in a quantum dot can lead to a power-law decay of electron spin coherence. It has also been observed that if, for example, the bath consists of only relatively few electron spins, the coherence of a central spin (in one particular case the spin of the so-called nitrogen vacancy center in diamond [2]) can also decay according to a power law rather than an exponential law.

1.3
Quantum Dots

For many practical applications of spins – for example to realize a quantum bit – solid state implementations are an attractive option. When electrons or holes – the mobile carriers of spin in semiconductors – are confined within a tiny structure with one or more dimensions smaller than the extent of the bulk wavefunctions, the electronic properties are drastically modified. When this confinement is along all three spatial directions, a "quantum dot" forms in which, like in a particle-in-a-box, as shown in Figure 1.2a, the energy levels of carriers are quantized. The band edges along one of the three spatial directions with resulting discrete energy levels for electrons and holes are depicted in Figure 1.2b. These energy levels can, following Pauli's exclusion principle, hold two electrons or holes of opposite spin direction. These "orbital states" can be filled sequentially starting from the lowest levels, the ground state of the quantum dot for each carrier species. Atomic physics holds an equivalent of this shell filling known as "Hund's rule". Due to these remarkable analogies to real atoms quantum dots are often referred to as their "artificial" counterparts [3]. Instead of continuous bands of conduction and valence band states, the energy eigenstates are now spatially localized within the dot, and separated by an energy that increases with increasing confinement.

When the temperature (T) is low enough such that $k_B T$ is smaller than the quantum dot energy level spacing (k_B is Boltzmann's constant), the quantized nature of the energy levels becomes apparent. For temperatures around 4 K, this requires a QD size of the order of 100 nm, consisting of $\sim 10^5$ atoms. In this regime, quantum dots will exhibit an atom-like spectrum of absorption and emission lines. Figure 1.3 shows a comparison of the emission lines of helium atoms and a semiconductor quantum dot. This correspondence between atoms and quantum dots provides a useful analogy, and the physics of quantum dots can be understood by borrowing ideas and concepts from atomic physics. For example, the optical pumping and control of quantum dot states has been demonstrated using the same tech-

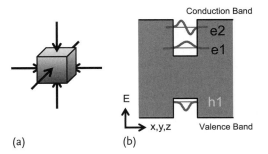

(a)   (b)

Fig. 1.2 (a) Schematic of a "quantum dot" in which carriers are confined in all three spatial direction in an area smaller than the de Broglie wavelength of the particle. (b) The confining potential, energy levels, and wavefunctions in a simple particle-in-a-box picture are illustrated for one spatial direction.

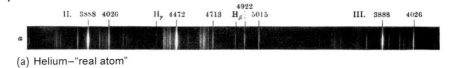

a

(a) Helium—"real atom"

s-shell

(b) Quantum dot—"artifical atom"

Fig. 1.3 Comparison of a part of the Helium atom spectrum (a) recorded on a photographic plate and a quantum dot spectrum (b) recorded using a CCD detector array. Figure reprinted with permission from [4]. Copyright by Wiley-VCH Verlag GmbH & Co. KGaA.

niques that were developed for controlling atomic states. On the other hand, the study of quantum dots offers much more flexibility than atomic systems. Many properties of quantum dots are tunable, including their size, shape, and material, which gives us a large degree of control. Even "artificial molecules" in which the interaction between two quantum dots can be switched "on" and "off" can be realized.

1.3.1
Spin-Based Quantum Information Processing with Artificial Atoms

In a classical computer, information is stored and processed in bits, each of which can take on one of two logic values. Once the values "0" and "1" have been assigned to the two eigenstates of a two-level system, such as a spin 1/2, the quantum mechanical spin dynamics can be viewed as the processing of information. A peculiarity of this information is that the binary values of a single bit can be brought into a coherent superposition. A bit with this property is called a quantum bit or qubit. According to the postulates of quantum mechanics, when a qubit is measured, it is always projected into one of its eigenstates, providing, for example a classical binary output after the end of a computation. In general, exploiting such uniquely quantum effects in spins or other two-level systems via unitary operations goes by the name of quantum information processing. This field can be divided into two categories: quantum computing, and quantum communication. These topics would already fill more than the space provided in this book. We only mention a few ideas here and refer the reader to the literature for more details.

Before touching on these two developments of quantum information processing, let us briefly consider a few particularities of qubits, namely, coherence and entanglement. We have seen above that coherence is basically the stability of the relative phase ϕ in Eq. (1.7) between the two eigenstates. Preservation of coherence is obviously a necessary condition for an undisturbed quantum computation. The decoherence time therefore imposes a limit on the minimal speed required for successful qubit operations. We already mentioned earlier that certain quantum effects are not captured in a semiclassical framework. A particularly important quantum property without classical counterpart is the entanglement of quantum states. Two spins are entangled if their total wavefunction cannot be written as a direct product of two single-spin states, such as $|\psi_1\rangle_1 \otimes |\psi_2\rangle_2$. Probably the most famous representatives of entangled states are the spin singlet,

$$|S\rangle = \frac{1}{\sqrt{2}} \left(|\downarrow\rangle_1 |\uparrow\rangle_2 - |\uparrow\rangle_1 |\downarrow\rangle_2 \right) \tag{1.16}$$

and the spin triplet with zero spin projection along the quantization axis z,

$$|T_0\rangle = \frac{1}{\sqrt{2}} \left(|\downarrow\rangle_1 |\uparrow\rangle_2 + |\uparrow\rangle_1 |\downarrow\rangle_2 \right). \tag{1.17}$$

Clearly, the remaining two triplet states, $|T_+\rangle = |\uparrow\rangle_1 |\uparrow\rangle_2$ and $|T_-\rangle = |\downarrow\rangle_1 |\downarrow\rangle_2$ factorize and are not entangled.

Quantum computing exploits the additional computational possibilities due to quantum mechanical complexity and parallelism in certain algorithms [5]. A famous example is the quantum algorithm by Peter Shor for the prime factorization of integers, which provides an enormous speed-up potential when factorizing large numbers. The crucial difference here lies in different scaling, as the time needed to factorize an integer on a classical computer grows exponentially with the number of digits $\log N$, while with Shor's algorithm it only scales polynomially [6]. The second famous algorithm is Lov Grover's quantum algorithm for search in an unstructured database [7].

David DiVincenzo has formulated criteria that need to be satisfied for quantum computation [8]. First, a suitable realization of a qubit must be found, in which information can be written, manipulated, and read out. Then, a register of qubits needs to be initialized at the beginning of a computation. The qubits must be sufficiently isolated from the environment to provide long enough decoherence times. In order to process quantum information, gate operations must be implemented. This requires high-precision control of single-qubit rotations and of switchable two-qubit interactions. It has been shown that single- and two-qubit operations are sufficient to implement any computation, that is, they form a universal set of gates [9, 10]. Finally the qubit register must be read out. The scalability is an additional criterion that needs to be met in practical implementations.

An electron spin in a quantum dot is a popular candidate for a qubit, since it is a natural two-state system. Electron spins in semiconductors have received much attention for quantum information applications because (i) semiconductor processing technology should make the scaling to large systems easier, (ii) electron spins in semiconductors have been found to have long coherence times relative to the expected times for gate operations, and (iii) spins and charge excitations can be initialized, addressed, manipulated, and read *both* by electrical and optical means. In recent years, a number of schemes have been demonstrated for achieving the requirements of state initialization, readout, and control for spin qubits; this is further discussed in later chapters. Nevertheless, there is still a long way to go before these elements can be put together in a functioning quantum computer. For a review on the recent status of spin-based quantum computing, see, for example, Cerletti *et al.* [11].

Quantum communication involves the transmission of quantum information from one place to another. This has applications, for example, in secure communication (cryptography), in teleportation of quantum states, and in superdense coding [12]. Cryptography involves sharing a secret key between two parties, such that the communication can be executed via a traditional communication channel, protected by a safe encryption. Quantum mechanics helps in distributing a secret key with the very issue we have when measuring a state: we project it into an eigenstate. For example, if polarization encoded photons are measured on the way by a third party in a different basis than the encryption scheme is using (which is the most probable case), the photon polarizations received in the end will just be randomized, which is detectable. Other schemes that have recently been implemented with quantum dot single-photon sources [13] involve the encoding of a key in a stream of single photons. Here, a third party would obviously quench the flow of information with a beam splitter attack, when detecting photons in a photon counter at the third party's site, which is also easily detectable in a communication scheme using control sequences. Quantum teleportation, in turn, consists of transmitting the quantum state of an object, for example a photon or an atom, faithfully onto a second object by performing a clever set of local measurements and by sharing a pair of entangled particles, for example two photons. Quantum teleportation has also been implemented using a single photon source provided by a quantum dot [14]. At this point we do not delve more deeply into these particular realizations.

Quantum communication necessitates a "flying qubit" – a carrier of quantum information that can be moved from place to place. Though spins and other qubit candidates such as single atoms can be moved over micron-scale distances within their coherence time, the only practical qubits for long-distance quantum communication are photon-based. Photons make ideal carriers of quantum information because they travel fast, and they have very long coherence times. The flip side to the long coherence time is that photons interact very weakly with each other – and a strong controllable interaction is a desired feature for quantum information processing. This leads us to consider a hybrid system with stationary qubits used for quantum information processing at either end, and flying qubits communicating faithfully between the two. This requires a way of converting stationary qubits to flying qubits. Fortunately, spins in optically active quantum dots couple to photons in a variety of ways, making this an intriguing platform for a potential hybrid quantum computing/communication system. In this book we want to introduce not only various ways how quantum information can be transferred between the photonic and spin domains but also how the electromagnetic wave nature of light can be used to coherently initialize, manipulate, and read-out the spin of an electron confined in a quantum dot.

1.3.2
Optically Active Quantum Dots

For a quantum dot to be considered "optically active" the interaction between light and carriers in the quantum dot must be sufficiently strong so as to make it useful for scientific investigations or technological applications. The definition of an "optically active" quantum dot is not entirely strict since virtually all quantum dots have some measurable interaction with electromagnetic radiation in the optical domain, that is, light. Nonetheless, here we will concern ourselves with those types of quantum dots that are primarily investigated by optical means. The confinement potential of these dots is typically such that they can hold *both* electrons and holes giving rise to a particularly strong interaction with light.

The primary way that light interacts with an optically active quantum dot is through transitions between the valence band and conduction band states in the quantum dot or in the surrounding semiconductor material. When light with a sufficiently small wavelength is incident on a quantum dot, transitions can be driven that serve to excite electrons from the valence band to the conduction band. Likewise, the inverse process occurs when an electron relaxes from a conduction band state to an unoccupied valence band state – light is emitted. This gives two measurable quantities for investigating the properties of a quantum dot: optical absorption and luminescence.

Further optical properties can also be observed and exploited in optically active quantum dots. Off-resonant interactions, such as the Faraday effect or Raman transitions are particularly useful for spin readout and manipulation, respectively.

1.3.3
"Natural" Quantum Dots

A very simple quantum dot system, which moreover exhibits an extremely high optical quality, are so-called "natural" quantum dots. The name originates from the fact that these dots form naturally in thin quantum wells. Such quantum wells are fabricated by deposition of two-dimensional films of semiconductor materials with different bandgaps. For example, if a thin layer (with a width $d \sim 7\,\mathrm{nm}$) of gallium arsenide (GaAs) is sandwiched between barriers made of aluminum gallium arsenide (AlGaAs) a potential well for both electrons and holes is formed perpendicular to the layers as shown in Figure 1.2b. The energy of the lowest level with respect to the band edges of the quantum well material (GaAs) depends on the width d of the well, that is, the GaAs layer. In the simplest approximation of a square well potential with infinite barriers we would expect a $\propto d^{-2}$ dependence of the ground state energy and, therefore, a wider well has a deeper lying ground state. As a matter of fact, semiconductor quantum wells are not always perfectly flat but exhibit monolayer-high steps. This situation is sketched schematically in Figure 1.4a. If the areas in which the well is thicker have the appropriate size (typically $\sim 100\,\mathrm{nm}$) monolayer fluctuations form quantum dots, which are therefore also sometimes referred to as "interface fluctuation quantum dots" (IFQD). These

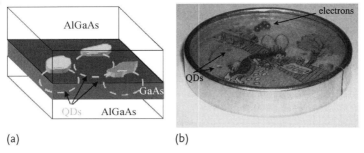

(a)   (b)

Fig. 1.4 (a) Illustration of interface fluctuation quantum dots: Quantum dots form at localized monolayer fluctuations of well thickness. (b) Toy with little balls and dimples.

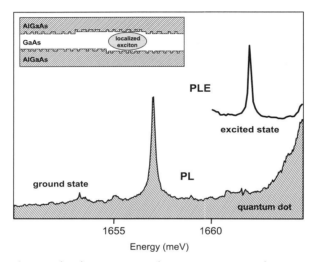

Fig. 1.5 Photoluminescence and PL excitation spectra of a "natural" quantum dot. The quantum dot geometry is shown schematically as an inset. Figure courtesy of W. Heller.

interface fluctuation quantum dots act like the little dimples in the toy shown in Figure 1.4b where little balls (representing electrons or holes) can be caught[1].

In the case of a GaAs-AlGaAs quantum dot the confined electrons and holes can recombine by emitting light, which is called photoluminescence (PL) if the electrons and holes were previously generated by light, for example by a laser. A typical example of a PL spectrum and a schematic of the quantum dot geometry is shown in Figure 1.5. This quantum dot shows a sharp atom-like PL line at and energy of 1657 meV. The observed linewidth is determined entirely by the resolution of

1) Since this toy does not know about spins, each dimple can only hold one ball. For deeper dimples higher occupancy states can be achieved.

the experimental apparatus, which underlines the high optical quality of the dot. In this figure the excited state absorption of the same quantum dot is also shown. It is measured in a PL excitation (PLE) experiment in which the intensity of the ground state PL signal is recorded as a function of the excitation laser energy. In such an experiment a high signal is only observed at laser energies at which the QD absorbs, which in this example occurs at an energy 5 meV higher than the ground state. Furthermore, on the right, at even higher energy, the onset of emission from the quantum well in which the dot is formed (peak at 1670 meV) can be seen.

The preceding discussions of quantum dots and of electron spin dynamics provide an introduction to these topics that will be delved into more deeply throughout this book. Chapters 3, 5, and 6 will treat the physics of spins in quantum dots in much more detail. Moreover, Chapters 7 and 8 describe a number of experimental observations of these phenomena. But first we will continue in Chapter 2 with different physical realizations of QDs. We will start by explaining the basics of semiconductor heteroepitaxy, which is the underlying method for the fabrication of embedded QD structures like "natural" dots, which we introduced in the previous section. We will also discuss the other prominent example of epitaxial quantum dots, the so-called self-assembled QDs after which we continue with a different technique to fabricate nanometer-size colloidal quantum dots.

2
Optically Active Quantum Dots: Single and Coupled Structures

In the rapidly growing field of quantum information processing a significant, but also quite challenging, goal is the realization of a scalable and robust hardware using a solid state system. Since the 1990s zero-dimensional semiconductor quantum dot nanostructures have attracted much attention because of their "atom-like" energy spectrum with the associated discrete density of electronic states. Moreover, the fabrication methods for many QD realizations provide the crucially required inherent scalability. After experiments in the last decade of the previous century demonstrated that carriers confined in QDs indeed are less affected by decoherence effects, charge, and in particular, spin excitations in these nanostructures were proposed as attractive qubits [11, 15–20].

Nowadays there are two main approaches used to fabricate QDs: (i) semiconductor heteroepitaxy and (ii) wet chemical synthesis. In this chapter we will introduce these fundamentally different techniques and describe the electronic structure of these dots based on the experimentally observed optical properties.

In contrast to "natural" and self-assembled QDs, which we will introduce in the first section as the most prominent examples for epitaxial QDs, colloidal QDs introduced in the second section of this chapter are not fabricated by epitaxial techniques but are synthesized from solution and, therefore, allow for shell structures providing an additional degree of freedom to engineer the electronic and optical properties of these nanostructures.

2.1
Epitaxial Quantum Dots

One of the first QD systems that was studied spectroscopically was the so-called "natural" or interface fluctuation QDs (IFQDs), which were already described briefly in the previous chapter. This simple but also high quality type of QD forms by monolayer fluctuations in thin quantum wells (QWs) [21–24]. In this QD system, experiments demonstrated the high quality and low susceptibility to optical decoherence processes of excitons despite the shallow confinement potential of these QDs [25–29].

Spins in Optically Active Quantum Dots. Concepts and Methods.
Oliver Gywat, Hubert J. Krenner, and Jesse Berezovsky
Copyright © 2010 WILEY-VCH Verlag GmbH & Co. KGaA, Weinheim
ISBN: 978-3-527-40806-1

Another widely investigated QD system is presented in the second section: These self-assembled QDs are nanometer-sized coherent islands grown by self-assembly in strained material systems via the so-called Stranski–Krastanow growth mode [30–34]. These islands show excellent structural and optical properties allowing for the observation of few-particle and coherent effects of excitons localized in individual nanostructures [35–45].

One remarkable breakthrough in the field of semiconductor fabrication techniques is the epitaxial deposition of different semiconducting materials layer by layer. The Greek root "epi-taxis" of the word epitaxy means "in the same manner", that is, preserving or adopting the structure of the underlying substrate. This versatile technique allows us to obtain novel structures in which carriers can be confined in one, two, or even all three spatial directions, known as quantum wells (QW), quantum wires (QWR), and quantum dots (QD), respectively. These systems led to the observation of novel physics like the quantum Hall effect, the quantum confined Stark effect, or the fabrication of materials with "artificial band structures" [46–49]; many of these topics are by now part of advanced student textbooks [50]. Moreover, advanced device concepts with functionalities not possible with conventional bulk semiconductor devices have been envisioned and realized. One famous example in this context is the quantum cascade laser [51]. These developments are often described by the term "band structure engineering" [52, 53].

In a reactor for semiconductor epitaxy the reactants of the material to deposit are brought to a (typically) heated substrate either in a vapor or as individual molecules. These source materials can be either in a chemically bound form, for example, a metal-organic compound in a carrier gas or as a molecular beam in an ultra-high vacuum chamber. In these two cases the layers on the growth interface form either by chemical vapor deposition or by forming layers of adatoms impinging on the surface. These two methods are called metal-organic chemical vapor deposition (MOCVD) and molecular beam epitaxy (MBE) and represent the two most frequently used techniques for the fabrication of high quality epitaxial semiconductor structures. In this book we will not describe these methods in much detail and refer interested readers to the extensive literature and textbooks on the wide field of epitaxial deposition techniques, for instance, [54–56].

In epitaxy one can distinguish between two types different growth: (i) homoepitaxy where the substrate and the deposited materials are the same and (ii) het-

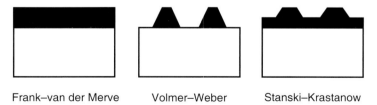

Frank–van der Merve   Volmer–Weber   Stanski–Krastanow

Fig. 2.1 Growth modes in semiconductor heteroepitaxy: Frank–van der Merve (2D), Volmer–Weber (3D), Stranski–Krastanow (2D–3D).

eroepitaxy where a different material is deposited on a substrate. Since in heteroepitaxy the deposited material and the substrate material are different the chemical and structural properties at the growth interface determine in which way the impinging material is deposited. In heteroepitaxy three different growth modes exist, which depend on the detailed energetic balance between surface and strain energy, depicted schematically in Figure 2.1. For almost identical lattice constants and crystal structures of the deposited material and the substrate either the Frank–van der Merve (FvdM) [57] or Volmer–Weber (VW) [58] growth mode takes place. The growth of the system in either of these growth modes depends on whether the sum of surface (α_2) and interface (β_{12}) energies is less or greater than the surface energy (α_1) of the substrate. FvdM growth mode is observed in material systems with $\alpha_2 + \beta_{12} < \alpha_1$, whereas VW growth mode occurs for $\alpha_2 + \beta_{12} > \alpha_1$. The most prominent material system with FvdM growth is Ga(Al)As whereas VW growth can be realized, for example, in the In(Ga)N/GaN material system.

In strained systems, like In(Ga)As/GaAs, In(Ga)As/InP, SiGe/Si or CdSe/ZnSe the Stranski–Krastanow (SK) growth mode occurs [30]. First, a highly strained two-dimensional wetting layer (WL) forms until, at a critical thickness, a transition to island growth occurs. During the WL formation, the elastic energy in this strained layer accumulates until at a critical thickness the crystal minimizes its total energy by formation of islands with new facets and edges. These lead to a reduction of the total free energy in this growth mode since the reduction of the elastic energy is larger than the increase of surface energy. The formation of new crystal facets increases the surface energy but this is overcompensated by the decrease of elastic energy. These islands grow coherently (i. e., without dislocations) until they reach a critical size and dislocation formation is the only way for strain relaxation to occur.

One major advantage of epitaxial QDs is that they can be grown in layers and, therefore, embedded in more complex structures containing, for example, contact

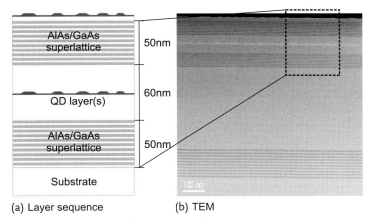

(a) Layer sequence   (b) TEM

Fig. 2.2 Schematic (a) and transmission electron micrograph (b) of a MBE grown heterostructure containing AlAs/GaAs superlattices and self-assembled QDs.

regions and other nanostructures. They can be used as an active emitting medium in high performance laser structures, optical modulators, emitters in photonic microcavities, or electrically coupled to doped layers. They have found widespread applications in novel optoelectronic devices with improved performance compared to systems of higher dimensionality. An example of an epitaxial layer sequence is shown schematically in Figure 2.2a. This sample contains AlAs/GaAs superlattices, which form in FvdM growth mode as well as In(Ga)As QDs grown by self-assembly in SK growth mode. In the transmission electron micrograph in Figure 2.2b the different materials can be clearly identified from the GaAs matrix as dark (AlAs) and bright (InGaAs) regions.

2.2
"Natural" Quantum Dots Revisited

As mentioned above, an MBE-grown heterostructure consisting of a layer of a semiconductor sandwiched between two layers of higher bandgap semiconducting material serves to confine electrons to the lower bandgap region and forms the familiar structure known as a quantum well. This provides a straightforward and flexible route to obtaining one of the required three dimensions of confinement needed to produce a quantum dot. In so-called "natural" or "interface fluctuation" quantum dots (already introduced in Chapter 1), the in-plane confinement arises from monolayer thick fluctuations in the quantum well width.

2.2.1
Structure and Fabrication

In ideal Frank–van der Merve heteroepitaxy, the interfaces between two materials are atomically flat. In reality however, the interfaces of a quantum well will have height fluctuations, increasing or decreasing by one atomic layer. As described in the previous chapter, these fluctuations result in a landscape in the in-plane direction of the quantum well with regions of varying thickness. Since the ground state energy in the conduction band of a quantum well increases as the well width decreases, an island in the quantum well surrounded by a region one atomic layer thinner forms an isolated potential minimum for a conduction band electron or valence band hole. For such an island of appropriate size, and at sufficiently low temperature, this potential minimum serves as a quantum dot.

Most commonly, a GaAs quantum well surrounded by AlGaAs barriers is used to form IFQDs providing confinement for electrons and holes. The growth is carried out in the same manner as a regular quantum well, except for a pause in the growth on the order of one minute at the two interfaces of the well. This pause allows the atoms at the surface to migrate around and form islands of the appropriate size (≈ 100 nm in diameter). Figure 2.3 shows an STM image of these fluctuations in an uncovered GaAs layer. Regions of different color represent steps in the height of plus or minus one atomic layer. From this micrograph one can directly see that

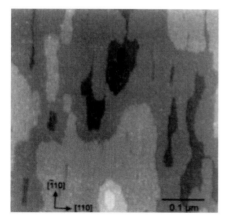

Fig. 2.3 STM image of a GaAs surface showing monolayer fluctuations from which IFQDs are formed. Regions of different shading indicate the monolayer steps on the surface. Reprinted with permission from [24]. Copyright (1996) by the American Physical Society.

these interface fluctuations are asymmetric and elongated along the $[\bar{1}10]$ crystal direction. This gives rise to an increased anisotropic exchange interaction between the electron and the hole forming the exciton. This interaction splits the ideally circularly polarized and thus spin-selective exciton transitions into a linearly polarized doublet [24]. For addressing spins in QDs via the polarization of light this has to be taken into account. A detailed discussion will be presented in Chapter 6. The planar density of QDs is typically quite high (on the order of $10 \, \mu m^{-2}$) and can be controlled to some degree [59] by adjusting the substrate temperature and waiting time before and after deposition of the GaAs layer. During these growth interruptions monolayer fluctuations form and a more homogeneous spatial distribution can be achieved. Since the formation does not rely on self-assembly or shell formation as is the case for Stranski–Krastanow and colloidal QDs, respectively, more complex structures like coupled pairs of IFQDs are extremely difficult to achieve. This fact limits the scaling for spin and exciton-based qubits to at most one or two qubits, respectively [29, 60].

2.2.2
Energy Levels and Optical Transitions

Under optical illumination with photon energy above the quantum well bandgap, electrons and holes can be excited into the conduction and valence bands of the quantum dots, respectively. The resulting photoluminescence (PL) spectrum (bottom of Figure 2.4b) shows two distinct broad peaks due to recombination of electrons and holes in the quantum well. The splitting between the two peaks observed originates from areas of different QW width of 10 and 11 monolayers with the 10 ML areas having a higher transition energy compared to the 11 ML ones. To

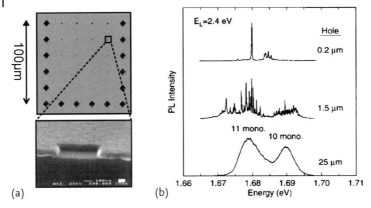

Fig. 2.4 (a) *Top:* Array of near-field optical apertures with sizes ranging between 200 and 800 nm in an aluminum mask. *Bottom:* SEM picture of a nominally 400 nm square aperture. (b) Photoluminescence spectra of GaAs IFQDs collected through apertures of various sizes. Reprinted with permission from [24]. Copyright (1996) by the American Physical Society.

confirm that the 11 ML thick regions are sufficiently small to form dots, individual interface fluctuations must be isolated. A convenient and frequently applied technique for the isolation of individual QDs are near-field shadow masks. Here a thin opaque metal layer (∼100 nm) is deposited on the sample surface. This mask is patterned with submicron apertures beneath which – in the optical near-field – a single QD or pair of QDs is located whilst the surrounding nanostructures are blanked out. Figure 2.4 shows PL spectra of IFQDs measured through increasingly small metal apertures fabricated atop the sample. As the aperture size is decreased, the broad low energy peak seen in the ensemble measurement breaks up into a number of sharp peaks corresponding to emission from a few individual QDs underneath the 1.5 μm aperture. With the smallest 200-nm-diameter apertures, which are in the same range as the average size of and distance between the IFQD determined by STM (see Figure 2.3), individual PL lines now appear. These are well-separated in energy, indicating the isolation of just one or very few QDs. This can be further confirmed by photon-photon time correlation measurements, which prove that these peaks come from a single quantum light emitter. By carefully adjusting the growth parameters, sufficiently large lateral interdot separations can be achieved, which then do not require the described masking technique [59].

Given the relatively weak confinement of an IFQD (typical potential barriers in the lateral direction of about 10 meV), these dots hold only up to two electrons and holes. Thus, there are only several possible configurations of charge in an IFQD. These different charge configurations can result in various lines seen in the PL spectrum of a single QD. For example, if a QD is occupied by one electron and one hole (the exciton state, often denoted X^0), the two can recombine, emitting a photon at the energy difference between the electron and hole states, minus the electron–hole binding energy. Likewise, if the QD is occupied by two electrons and one hole, or two holes and one electron (negatively or positively charged excitons, denoted X^- or X^+ respectively), luminescence will be observed at the energy difference between the charged exciton state and the single electron or hole state. Finally, a fourth PL line is often observed due to recombination from the biexciton state ($2X^0$), leaving behind a neutral exciton in the QD. The different species of excitons in QDs and the energy spacing between their emission lines will be discussed later in Section 2.3.2 for the example of self-assembled QDs. Unlike the case of self-assembled In(Ga)As QDs discussed next, the confinement in IFQDs is sufficiently weak that higher-charge states or shells are typically not confined in the dot. This makes this type of QD unsuitable for high-temperature applications; however, fundamental proof of principle experiments have often been demonstrated first in this system.

In addition, by embedding a layer of quantum dots in a diode, the charge state of the QD can be tuned with an applied voltage across the device. From this type of data, one can determine the energies of the various charge configurations within the QD. These types of devices will be discussed in detail in Chapter 4 and they are the basis for many of the experimental results presented in Chapters 7 and 8.

2.3
Self-Assembled Quantum Dots

The fabrication of self-assembled QDs is based on the Stranski–Krastanow growth mode that we introduced in Section 2.1. When In(Ga)As is deposited on a GaAs substrate the $\sim$ 7% mismatch between the bulk lattice constants of the two materials leads to the self-assembly of InAs islands on a two-dimensional wetting layer when the InAs has reached a critical coverage of $\sim$ 1.7 ML [33, 34]. Figure 2.5 shows an atomic force micrograph (AFM) of an ensemble of uncapped InAs islands grown on GaAs. These islands are randomly distributed on the surface and have a height of 5–10 nm and a diameter of 20–30 nm. This size distribution has a strong impact on the optical emission of ensembles of these QDs.

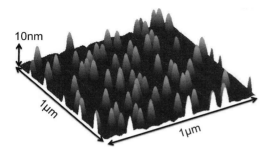

10nm

1μm

1μm

Fig. 2.5 Atomic force micrograph of an ensemble of InAs islands on GaAs. Micrograph courtesy of P. M. Petroff, T. A. Truong and H. Kim, UC Santa Barbara.

For broad size distributions a strong inhomogeneous broadening of the emission reflects the resulting variation of the electronic levels in the different dots in the probed ensemble. During the two-dimensional wetting layer growth only a certain fraction ($\sim 20\%$) of the deposited indium is incorporated into the WL. The excess material is present at the surface. The In concentration increases until the critical value for island nucleation is reached. Afterwards, an efficient transfer of In adatoms from the surrounding material into the island takes place leading to an increased In concentration at the apex and an "inverted pyramidal" alloy profile [61, 62].

For studies of individual QDs or QD pairs, material with a sufficiently low surface density of dots is desirable. An elegant approach to fabricate areas of different surface densities on a single wafer can be achieved by MBE. Such a gradient in the surface density can be obtained by exploiting the fact that in MBE a molecular beam impinges from a relatively small source onto the substrate. This molecular beam is not homogeneous over the area of the substrate but typically forms a gradient across the wafer. For the fabrication of homogeneous films, as for quantum wells, this material gradient has to be compensated for by rotating the substrate holder. However, when the substrate rotation is intentionally stopped during the deposition of the QD material (for instance, InAs) the thickness of this film varies and decreases from the side close to the In cell to the opposite side. The nominal amount of material can be set to a value that in the high coverage area the critical thickness for island formation is reached, that is Stranski–Krastanow growth of island occurs whilst on the opposite side the deposited material is no longer sufficient and only a wetting layer is formed [34]. When moving along the material gradient across this nucleation border the QD surface density decreases from several 100 per μm^2 to 0. In the transition region, which can, depending on the MBE chamber's geometry, extend over several millimeters, individual QDs can be isolated directly by diffraction limited microphotoluminescence or in combination with the previously discussed masking technique.

2.3.1
Strain-Driven Self-Alignment

Since the formation of self-assembled QDs is driven by the strain present in the material, this property can be used to fabricate self-aligned multilayer structures [63]. For an overview on recent advances on strain-driven self-alignment and self-assembly of coupled nanostructures on planar and patterned substrates we refer the interested reader to recent literature [64–66]. A combination of the different approaches together with optical cavities might provide a direct route to a controllable architecture for QD-based quantum information processing.

The mechanism for the formation of vertically aligned pairs or even chains of QDs is schematically depicted in Figure 2.6. If a seed layer of QDs is overgrown by a thin spacer layer the strain field is not fully relaxed at the surface but modulated by the overgrown islands. This leads to preferential nucleation sites for islands in the second layer right on top of the overgrown one. This procedure can be repeated

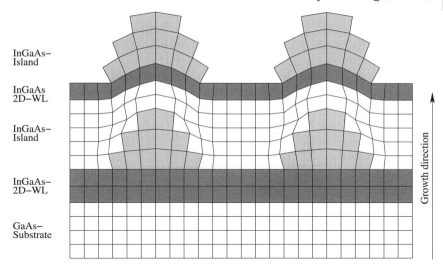

InGaAs–
Island

InGaAs
2D–WL

InGaAs–
Island

InGaAs–
2D–WL

GaAs–
Substrate

Growth direction

Fig. 2.6 Strain-driven self-alignment of self-assembled QDs.

several times leading to the formation of vertically aligned columns of QDs. For thin spacer layers (<15 nm) a stacking probability close to unity can be achieved as demonstrated in Figure 2.7a [63]. The scheme can be extended to multiple stacks of QDs. In Figure 2.7b we present an STM image taken on the cross-section of a cleaved sample. In the image five stacked InAs QDs can be identified. Since the dots in each layer are grown under nominally identical conditions, and each dot experiences more strain than the one below it, the lateral dimension of the islands increases from the bottom to the top layer [67]. By adjusting the growth parameters in the stacked QD layer(s) the relative size of the dots can be tailored [68–71]. This method is simpler than other techniques such as twofold cleaved-edge overgrowth (CEO), a technique that will be described briefly later, since it uses an intrinsic property of strained heteroepitaxy to achieve ordering. However, due to the inhomogeneous nature of the island formation the atomistic precision offered by CEO cannot be achieved [72]. After the first realization of strain-driven self-alignment in the InGaAs material system, such structures were investigated using optical and electrical spectroscopy in order to explore electronic coupling effects in these nanostructures. In experiments studying individual pairs of stacked QDs, indirect evidence for quantum dot coupling was obtained [73–76]. More recently, a clearer, unambiguous demonstration of coherent coupling has been observed [77]. In the context of quantum information processing, excitons in electronically coupled QDs were found to show long decoherence times, much longer than the timescales required for coherent manipulation using picosecond laser pulses [17, 78].

This method can be further expanded for the fabrication of columnar structures, so-called quantum posts [79, 80]. The fabrication relies on a deposition cycle of alternating layers of InAs and GaAs where after each layer the growth is interrupted for approximately one minute. The number of repetitions of this cycle determines

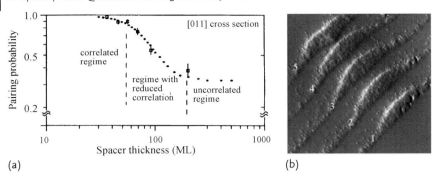

(a)   (b)

Fig. 2.7 (a) The pairing probability is found close to unity for spacers <15 nm. Reprinted with permission from [63]. Copyright (1995) by the American Physical Society. (b) X-STM of a fivefold stack of InAs QDs demonstrates perfect alignment of the islands. Reprinted with permission from [67]. Copyright (2003) by the American Institute of Physics.

the height of the nanostructure, which can exceed 60 nm. Furthermore, the diameter and In content can be controlled along the QD axis, that is the growth direction. Such quantum posts are a recent example of ongoing development in the field of self-assembled nanostructures with tailored structural and optical properties.

2.3.2
Optical Properties and QD Shell Structure

Photoluminescence (PL) spectroscopy is a powerful and widely applied technique to investigate the level structure of QDs. In this type of experiment, electron–hole pairs are photogenerated above the bandgap of the matrix material and then relax over a sub-ps-timescale into the QDs. Carriers occupying the confined levels can then recombine radiatively with typical lifetimes of $\sim$1 ns, much longer than the relaxation time to the s (ground) states for electrons and holes. The number of carriers present in the system can be adjusted by the optical excitation density. An example is shown in Figure 2.8, where an ensemble of $\sim 10^7$ QDs is probed by PL for two levels of excitation density. For low excitation density ($P_0 = 0.1\,\mathrm{W}\;\mathrm{cm}^{-1}$, black line) each QD captures occasionally a single electron-hole pair on average. Since the relaxation times to the s-states are much shorter than the radiative lifetime, mainly emission from the s shell is observed at low excitation densities. As the excitation density is increased by two orders of magnitude (gray line), more carriers are present in the system and the higher shells (p, d, f) are subsequently filled. This phenomenon is known as level-filling and represents typical behavior for In(Ga)As QDs. As we discuss in more detail in Chapter 5, an electron in a p shell will recombine with a hole in the p shell, a d electron with a d hole, and so on, due to the optical selection rules. The emission of the higher shells is detected in the PL spectra at higher power densities. Although these different transition lines can often be clearly resolved in ensemble measurements, as for the sample shown

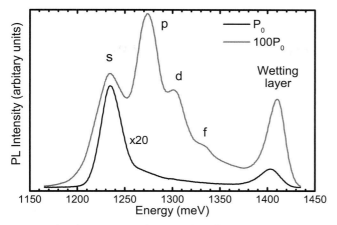

Fig. 2.8 Photoluminescence of an ensemble of highly homogeneous InGaAs QDs. Four shells are clearly resolved and a characteristic level-filling is observed as the excitation power is increased.

in Figure 2.8, the PL peaks are typically still inhomogeneously broadened due to size and morphology fluctuations in the quantum dot ensemble. In such ensemble measurements, usually none of the more delicate features such as the exciton fine structure, as discussed in Section 6.2, can be resolved. Nevertheless, PL spectroscopy performed on ensembles of self-assembled QDs is a powerful method to characterize the optical properties of these nanostructures.

In order to describe the interband optical transitions of QDs, the confined states of electrons and holes have to be modeled in a realistic way. Already in the early years of QD research, it was found that a simple harmonic oscillator model provides a fairly good agreement with the experimental spectra. We introduce this basic model here and compare the observed optical properties of self-assembled QDs to its predictions. In InAs/GaAs as well as GaAs/AlGaAs QDs an attractive confinement potential exists for both electrons ($\alpha = e$) *and* holes ($\alpha = h$). This is usually referred to as a Type-I band alignment and obviously facilitates optical interband transitions of these QDs[2]. The discussion presented here is more general and also holds for most other QD systems.

The typical height of self-assembled QDs in the growth direction z is small (≤ 5 nm). The confinement along z is therefore treated as for a narrow quantum well. Since the quantized sub-bands of a narrow quantum well are split far in energy, only the ground state sub-band E_z^α is relevant for the experimentally observable dynamics of electrons and holes, respectively. In contrast, the lateral dimension of the QDs in the xy plane is typically much larger ($\gtrsim 15$ nm), providing an aspect ratio of $\lesssim 1 : 3$. The in-plane confinement is therefore much weaker than

2) In Type-II band alignment, in contrast, either electrons or holes are confined in the dot, while the carriers with opposite charge are being pushed out of the dot. A detailed discussion can be found in most textbooks on semiconductors, for example, [50, 56].

along z and determines the observable shell structure of the QD. The in-plane confinement for electrons and holes is given approximately by a radially symmetric, two-dimensional harmonic potential [81–83].

This approximation can be justified by the observed equidistant shells in the QD spectrum like the typical example shown in Figure 2.8. The strength of the confinement for the two carrier species is given by the frequencies ω_α. The eigenenergies of this well-known problem are given by

$$E^\alpha_{m,n} = \hbar\omega_\alpha(m + n + 1) \tag{2.1}$$

with the associated two quantum numbers n and m. From these quantum numbers $m, n = 0, 1, 2 \ldots$, we can directly obtain the level scheme of such an ideal model QD. The z-component of the angular momentum for each level with quantum numbers m and n is obtained from the formula $L^\alpha_{mn} = \pm(m - n)$, where the "$+$" and "$-$" signs apply for electrons and holes, respectively [81]. In analogy to atomic physics [3], levels with their quantum numbers adding up to $m+n = 0$, $m+n = 1$, $m + n = 2, \ldots$ are labeled $s, p, d, \ldots$ shells, respectively, in Figure 2.9.

In Figure 2.9b the allowed interband transitions are depicted schematically by vertical arrows. We study these transitions more closely in Chapter 5 and see there that they are electric dipole transitions. Here, let us just have a brief look at the transition energies. As a rough approximation, consider for the moment noninteracting electrons and holes in the quantum dot. The transition energy for an electron–hole recombination is given in this situation by

$$E_{\text{transition}} = E_g + \sum_{\alpha=e,h} E^\alpha_z + \sum_{\alpha=e,h} E^\alpha_{m_\alpha,n_\alpha} = E_{g,\text{eff}} + \sum_{\alpha=e,h} E^\alpha_{m_\alpha,n_\alpha} , \tag{2.2}$$

where the usual bandgap energy E_g of the material, including strain, and the confinement energy in z direction can be added to an effective bandgap energy, $E_{g,\text{eff}} = E_g + E^e_z + E^h_z$. The two-dimensional harmonic oscillator energies $E^\alpha_{m_\alpha,n_\alpha}$ with quan-

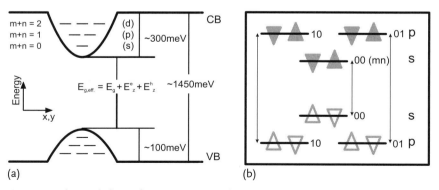

(a) (b)

Fig. 2.9 (a) The parabolic confinement potential of a QD leads to the formation of equally spaced energy levels labeled s, p, d for electrons and holes. (b) Structure of the s- and p-shell with the allowed interband transitions indicated by vertical arrows.

tum numbers n_α and m_α describe the in-plane energies of electrons and holes. Optical transitions among these levels give rise to the characteristic lines associated with the quantum dot transitions s, p, d,... shown in Figure 2.8. Later in Section 3.4 we show that the Coulomb interaction additionally induces characteristic shifts in the PL spectra, which we have neglected here for simplicity.

2.4
Alternative Epitaxial Quantum Dot Systems

In addition to "natural" and self-assembled QDs, which we introduced in the previous sections, many other approaches to achieve quasi-zero-dimensional structures have been followed over the past years. Some of these systems can be addressed and controlled optically, while others cannot, depending mainly on the fact whether or not *both* carrier species are confined in the dot.

2.4.1
Electrically Gated Quantum Dots

The most prominent example of a QD system that only provides confinement for one carrier species – in almost every case electrons – are QDs that are induced in a two-dimensional electron gas (2DEG). Such a 2DEG forms at the interface between two semiconductors with different band gaps, for example AlGaAs and GaAs. When a thin doped layer (modulation doping) is incorporated close to the hetero-interface in the high bandgap material (AlGaAs) the carriers are transferred to the low bandgap material (GaAs), and due to the electrostatics (positive donor atoms in the AlGaAs and electrons in the GaAs) a triangular potential for electrons forms at the interface giving rise to a two-dimensional channel in which electron transport can take place. A similar situation exists at the oxide-semiconductor interface in metal-oxide-semiconductor field effect transistors. Such hetero-interfaces have found direct application in devices such as high-electron-mobility transistors (HEMT) or modulation doped field effect transistors (MODFET). In addition to QDs artificially defined in these high mobility 2DEGs another area of basic research in these systems is the study of the integer and fractional quantum Hall effects. The latter was indeed observed for the first time in a sample with a 2DEG at an AlGaAs/GaAs hetero-interface [47]. A detailed discussion of the 2DEG and its physics and application can be found in most advanced student textbooks on low-dimensional semiconductor structures, for example [50].

When metal gate electrodes are defined on the surface of a sample containing a 2DEG the voltage applied to these gates can be used to repel or deplete the 2DEG electrons locally right underneath the electrode. Thus, for a 2DEG near the sample surface and for nanometer-sized gates suitable contact geometries, which isolate little puddles of electrons that are isolated from the surrounding 2DEG, can be designed. Such an example is shown schematically in Figure 2.10a. Here the non-depleted regions in the 2DEG are shown in gray. A QD is formed on the left side

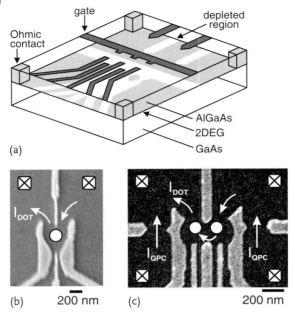

Fig. 2.10 Gate defined QDs in a 2DEG. (a) Schematic of a double-dot structure the 2DEG (gray shaded region) is depleted locally underneath the gate electrons thus defining and isolating QDs in between. (b) and (c) SEM images of a single-dot and a double-dot structure, respectively. Reprinted with permission from [20]. Copyright (2007) by the American Physical Society.

between two of the long finger gates, and the shorter gate between then can be used to tune the number of electrons in the QD. Figures 2.10b and 2.10c show scanning electron micrographs of gate-defined structures with a single-dot and a double-dot structure, respectively. In Figure 2.10c two so-called quantum point contacts are located next to the two dots, which can be used as extremely sensitive probes for the number of charges on the adjacent QD. Such gate-defined QDs are widely studied for implementation of spin-based quantum computation schemes. However this can be typically done only by electrical means since no confinement potential for holes is formed at the heterojunction. Moreover, their confinement is extremely weak, which typically requires that experiments have to be performed at very low temperatures of tens of milli-Kelvin in dilution refrigerators. Moreover, since only one carrier species is confined, semiconductors with indirect bandgap in k-space can be used. In this context Si/SiGe is an interesting material system since the nuclear environment can be engineered. Recently, significant progress has been made in the fabrication of isotopically engineered QDs and channels made from nuclear-spin-free ^{28}Si [84, 85].

A detailed discussion of this very active field of research would go beyond the scope of this book. We refer the interested reader to recent review articles on these types of QDs (for instance, [20, 86]) where exciting results like coherent control of single electron spins and conditional operations using coupled spin or direct electron spin resonance are presented.

A related two-dimensional system that attracted wide-spread attraction recently is graphene. In these single layers of sp^2 hybridized carbon, QDs can be defined, for example, by etching a nanometer-sized region. Due to the virtually nuclear spin free environment, this material has been proposed for spin-based quantum computation [87–89].

2.4.2
Advanced MBE Techniques

Alternative approaches using advanced epitaxial techniques such as the growth of pyramidal QD structures or twofold cleaved-edge overgrowth (CEO) have produced QDs of high optical quality. In these approaches intersections between QWs are used to form a QD. At such an intersection, analogous to a IFQD, the carrier wavefunctions can spread out over a larger volume, therefore lowering locally the effective bandgap of the system. Using this principle a QD can be realized at the intersection point of three QWs [72, 90, 91].

This principle is depicted schematically in Figure 2.11a for CEO. In this technique the sample is fabricated by a three-step process where in the first step a multi-QW structure (MQW_1) is grown. In the two following steps the sample is cleaved

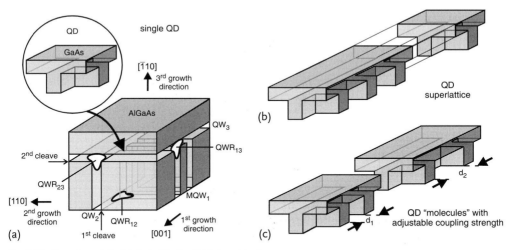

Fig. 2.11 Quantum dots fabricated by twofold cleaved edge overgrowth. (a) A single dot forms at the intersection between three QWs. (b + c) By using a multi-QW structure or superlattice a chain of QDs can be realized with extremely high precision and the interdot coupling strength can be adjusted by their separation. Reprinted from [91] with kind permission of Springer Science+Business Media.

inside the MBE in ultrahigh vacuum and the "fresh" facet is overgrown by another QW perpendicular to the already existing structure. At the intersection between all three wells a dot is formed as the wavefunctions of electrons and holes can extend over a larger volume. By appropriate design of the MQW_1 sequence isolated QDs, pairs or even long chains of QDs can be realized as shown in Figure 2.11b and 2.11c. Due to the atomic precision offered by MBE, QDs can be defined and positioned with monolayer accuracy. This led the precision of MBE. This led to the first and, for a long time, only observation of coherent coupling effects in an isolated QD "artificial molecule" [72]. In contrast to "natural" and self-assembled QDs, where electrical contacts and gates can be realized in a straightforward way, as we will show in Chapter 4, this combination of optical and electrical access turned out to be much more challenging for CEO devices.

Another method using the same concept of "providing more space for the wavefunctions to spread out and lower the state energy" are pyramidal QDs, which are fabricated by MOCVD. Here a single order multi-QW structure is grown inside pyramids etched into the wafer surface. The QWs are then grown on the sidewalls of these pyramids. However at the edges their thickness is slightly increased. Therefore, a high quality, optically active QD or pairs of QDs form at the tip of this pyramid [92, 93] and many interesting experiments have been performed using these types of QDs [94].

CEO and pyramidal QDs represent only two of many promising examples that exploit advanced epitaxial techniques. Many more elegant approaches exist to obtain QDs, coupled QDs, or even QD crystals using such (quasi-) planar methods. These are discussed in detail in recent books and reviews, for example, [64, 65].

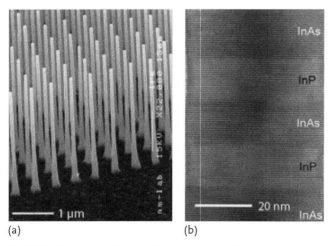

(a)   (b)

Fig. 2.12 (a) An array of prepositioned nanowires. Reprinted from publication [96] with permission from Elsevier. (b) TEM image of an InAs nanowire in which a QD-like structure is isolated by two InP barriers (light gray regions). Reprinted with permission from [97]. Copyright (2004) by the American Chemical Society.

2.4.3
Nanowire Quantum Dots

All techniques presented until now produced more or less planar structures in which QDs are embedded. A completely different system are nanowires, which are fabricated by a so-called bottom-up technique [95–98]. Here typically a catalyst or nucleation center is used to induce the growth of tiny needles with diameters down to a few nanometers and lengths that can exceed several microns. The material for these nanowires is provided either by a molecular beam in an MBE chamber or by metal-organic or other gaseous compounds in a chemical vapor deposition reactor. Such nanowires can be grown in a wide range of material systems including the group IV like silicon and germanium and group III–V compound semiconductors such as InGaAs and InGaP. Figure 2.12a shows an SEM image of an array of prepositioned nanowires. These nanostructures have attracted significant interest since they present an isolated and inherently one-dimensional system. Moreover, since they can be fabricated by established epitaxial techniques heterostructures can be obtained *within* the nanowire. By using the well known materials one can design and fabricate optically active QDs by embedding a thin layer of material with a lower bandgap inside the nanowire. An example of such a junction is shown in Figure 2.12b where two InP barriers (light regions) isolate a thin layer of InAs, which then can form a QD. In addition, the fabrication of multi-QD structures can be directly achieved. These nanowire QDs can not only be probed optically but also electrically providing additional degrees of freedom to investigate the electronic properties of these QDs [99, 100].

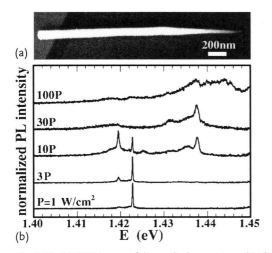

Fig. 2.13 (a) SEM image of the studied nanowires. (b) Sharp-line PL of a single nanowire quantum dot as a function of excitation power. Reprinted with permission from [99]. Copyright (2003) by the American Institute of Physics.

An example of a sharp PL line of an InAs QD embedded in a GaAs nanowire is shown in Figure 2.13b. An SEM image of one of the nanowires studied is shown in 2.13a. In the power dependent PL (the excitation increases from the lower to the upper spectrum in 2.13b) the filling of shells is observed comparable to self-assembled QDs.

We have noted before that nanowires can be also probed electrically making them a very versatile system. Moreover, for electrical spectroscopy gate electrodes can be also used to define a QD without the need for heterojunctions within the wire. This method, known from carbon nanotubes [101, 102], can also be successfully applied to nanowires, underlining the potential for applications in electronics and photonics in this system [103].

2.5
Chemically-Synthesized Quantum Dots

An alternative route to quantum dot fabrication relies on wet chemical synthesis [104, 105]. The nucleation and subsequent growth of semiconductor nanocrystals in solution provides a simple and highly flexible method for generating large ensembles of quantum dots.

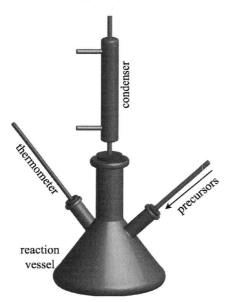

Fig. 2.14 Apparatus for synthesis of colloidal nanocrystals. Precursor molecules are added to the coordinating solvent through one neck of the flask; the condenser prevents evaporation of the solvent, and a thermometer measures the temperature.

2.5.1
Colloidal Growth

A typical setup for growth of nanocrystal QDs (NCQDs) is shown in Figure 2.14. A number of different growth recipes have been developed, but they all follow the same basic outline of colloidal growth [105]. A coordinating solvent is prepared in the flask and heated to some high temperature (around 300 °C). Another solution containing metal-organic precursor molecules (e. g., dimethyl cadmium, Me_2Cd, and trioctylphosphine selenide, TOPSe, to produce CdSe nanocrystals) is then quickly injected into the flask. With a high enough concentration of precursors, nanocrystals will begin to nucleate. As the nucleation continues, the remaining concentration of precursors decreases, and the temperature drops. Fairly rapidly, a threshold is crossed beyond which nucleation of new nanocrystals stops. The temperature is then held at some lower value (around 200 °C), which allows the nuclei to continue to grow. Once the nanocrystals have reached the desired size, the temperature is lowered further, and the growth is arrested. As long as the time during which nucleation occurs is short compared to the growth time, the ensemble of nanocrystals will be fairly uniform in size. In fact, typical synthesis procedures yield nanocrystals with a size uniformity of ±5%, which corresponds to an accuracy of about ±1 atomic layer for typical NCQDs. High-resolution TEM imaging (Figure 2.15a) of the resulting crystals shows good crystalline order, though given the large ratio of surface to interior atoms, the lattice is significantly strained. Most nanocrystal QDs are roughly spherical in shape, though there is often some faceting of the surfaces reflecting the underlying crystal structure.

Variations on this basic procedure provide great flexibility in the type of nanocrystals produced. Most commonly, II–VI semiconducting materials have been used to make nanocrystal quantum dots, such as cadmium, zinc, or mercury from the II column and sulfur, selenium, or tellurium from the VI column. However, nanocrystals can be fabricated from virtually all types of semiconductors, including

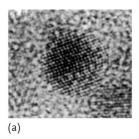
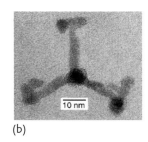
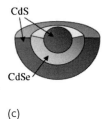

(a) (b) (c)

Fig. 2.15 (a) High-resolution TEM image of a single CdSe nanocrystal. Reprinted with permission from [105]. Copyright (1993) by the American Chemical Society. (b) A branched tetrapod nanocrystal. Reprinted with permission from [107]. Copyright (2000) by the American Chemical Society. (c) Cut-away schematic of a layered "quantum dot-quantum well".

III–V materials such as InAs, oxides such as ZnO, and group IV semiconductors such as germanium and silicon [106].

Additionally, one can also control the shape of the nanocrystals by changing the growth parameters [107]. Elongated nanocrystals can form "quantum rods", or even more fanciful creatures such as branched tetrapods, shown in Figure 2.15b. Another interesting degree of freedom afforded by colloidal nanocrystal growth is the ability to layer different materials within a single nanocrystal [29, 108, 109]. This can be accomplished by following essentially the same growth procedure outlined above to nucleate nanocrystals, then repeatedly exchanging the solvent to alternate between the desired cations and anions. This process results in an onion-like crystal with concentric layers of different materials, built up one atomic layer at a time. By building a structure of multiple semiconductors with different bandgaps, one can effectively tailor the potential felt by electrons and holes within a nanocrystal quantum dot. This technique is often used to grow a higher bandgap capping layer around a lower-bandgap core, to help isolate electrons and holes from the environment. Alternatively, more complex layered structures can be formed such as a "quantum dot-quantum well" (or "quantum shell") in which electrons and holes are confined to a spherical shell of a low-bandgap material sandwiched between a higher bandgap core and outer shell (see Figure 2.15c) [110–112].

An interesting feature of nanocrystal quantum dots, as compared with other types of dots, is that once the fabrication is complete, the story is not over. Many properties of the dots can still be affected by engineering their environment. The dots can be removed from solution to form a close-packed solid, they can be dispersed on a surface, they can be chemically functionalized, they can be embedded in a polymer, they can be arranged in an array on the surface of a virus – the list is essentially endless.

2.5.2
Energy Level Structure and Optical Properties

The spectrum of energy levels in a nanocrystal quantum dot is tricky to calculate. There are two approaches: to treat the nanocrystal as a small piece of bulk material using the effective mass approximation, essentially solving the Schrödinger equation with the appropriate boundary conditions, or to treat it as a collection of individual atoms, finding the energy level spectrum using, for example, density functional theory. Neither of these approaches is ideal. It is unclear whether the assumptions of a bulk material still apply to a nanocrystal, which may just contain a few hundred atoms. On the other hand even with just a few hundred atoms, accurate first-principles calculations become difficult. Nevertheless, both approaches have yielded results that are consistent with experimental evidence. Even in structures containing layers just a few atoms thick, the effective mass approximation has proved fairly successful in predicting the energy levels.

Following the first approach described in the preceding paragraph, one finds spin-1/2 electron states analogous to the energy levels of a hydrogen atom [113] (see also Section 3.3). The wavefunctions are composed of the underlying Bloch wavefunctions modulated by an envelope function, which can be labeled nL_e, where n is the principle quantum number, and L is S, P, D, and so on, indicating the angular momentum quantum number of the envelope function. For example, $1S_e$ denotes the lowest energy conduction band state with an envelope function with angular momentum quantum number $l = 0$. For the hole states, one must include the orbital angular momentum of the underlying valence band states. These states can be labeled nL_f where n is the principle quantum number, f is the total angular momentum of the hole state, and L is the lowest angular momentum component of the envelope function (the hole state envelope functions have a component of angular momentum l and $l + 2$). For example, the state $2P_{3/2}$ denotes the second highest energy (second closest to the top of the band) valence band state with an envelope having a component with $l = 1$ and $l = 3$, and with total angular momentum $3/2$. That is, this hole state contains states from the heavy-hole and the light-hole band, forming with the orbital angular momentum states a total angular momentum $3/2$ state.

Figure 2.16a shows the calculated and measured low-lying electron and hole states in a CdSe nanocrystal as a function of dot diameter. Note that the spacing between the low-lying energy levels can be quite large – on the order of 100 meV. Since this energy is large compared to $k_B T$ at room temperature ($\sim$25 meV), semiconductor nanocrystals behave nicely as quantum dots even at room temperature. The lowest energy electron and hole states in a typical nanocrystal quantum dot are the $1S_e$ and $1S_{3/2}$ states, respectively. The $1S_e$ state is twofold degenerate due to the electron's spin $1/2$, and the $1S_{3/2}$ state is fourfold degenerate, corresponding to the projections of angular momentum $J_z = -3/2, -1/2, 1/2, 3/2$. Therefore, if an electron–hole pair (i. e., an exciton) is optically excited in the dot, its lowest energy state will be eightfold degenerate.

Additional considerations may be taken into account when calculating the spectrum of states in a nanocrystal quantum dot [113]. For example, deviations from a spherical shape, or crystal anisotropy may be included, as well as the electron–hole exchange interaction. These considerations mix various electron and hole states, and lift the degeneracy mentioned above. This splitting will depend rather sensitively on nanocrystal size and shape. Figure 2.16b shows the calculated splitting of the $1S_e$ and $1S_{3/2}$ states as a function of dot diameter using an experimentally determined nanocrystal shape distribution. The different states are labeled by the projection of the exciton's total angular momentum, with the solid curves corresponding to total angular momentum $J = 1$ and the dashed curves to $J = 2$.

As usual, the optical transitions in nanocrystal quantum dots are governed by selection rules requiring conservation of angular momentum. That is, the absorption or emission of a circulary polarized photon must be accompanied by a change in the total angular momentum by ± 1. The optically active transitions are indicated in Figure 2.16b by the solid lines, and the forbidden transitions are indicated by dashed lines. It is interesting to note that the lowest energy exciton state is not an optically allowed transition. Thus electrons and holes can only be excited into higher energy states, and once the exciton has relaxed into its ground state, it cannot recombine optically (at least to first order).

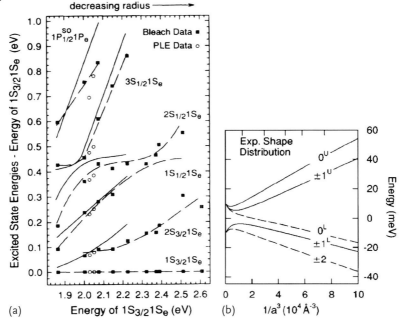

Fig. 2.16 (a) Calculated and measured spectrum of size levels in CdSe NCQDs vs. nanocrystal size. Data points are from experiment, solid lines are from theory, dashed lines are guides to the eye. Reprinted with permission from [114]. Copyright (1994) by the American Physical Society. (b) Calculated splitting of the $1S_e - 1S_{3/2}$ transition energy due to various perturbations mentioned in the text, using an experimentally determined distribution of nanocrystal shapes. Reprinted with permission from [113]. Copyright (1996) by the American Physical Society.

Figure 2.17 shows photoluminescence (PL) spectra (dashed lines) and optical absorption spectra (solid lines) for various sizes of CdSe nanocrystal quantum dots. As expected from "particle-in-a-box" considerations, and as borne out by the calculation shown in Figure 2.16a, the measured spectra shift to higher energy as the nanocrystal size decreases. The PL and the lowest energy peak in the absorption spectrum arise from the $1S_e$–$1S_{3/2}$ transition. As mentioned above, the lowest energy transition with significant oscillator strength is not the exciton ground state. This explains the ("Stokes") shift between the lowest energy peak in the absorption and the PL. Excitons are generated in the higher energy, optically active states and then rapidly relax into the ground state. Even though direct recombination is forbidden from this state, higher order processes involving phonons or defects can eventually allow the emission of a photon. This presence of this "dark" exciton ground state is confirmed by observations of very long radiative lifetimes for excitons in nanocrystal quantum dots (on the order of microseconds). For comparison, the radiative lifetime of self-assembled InAs dots is on the order of nanoseconds.

In many cases, the $1S_e$–$1S_{3/2}$ exciton states are essentially the only relevant states. Electrons and holes excited into higher energy levels relax quite rapidly ($\sim$ tens of

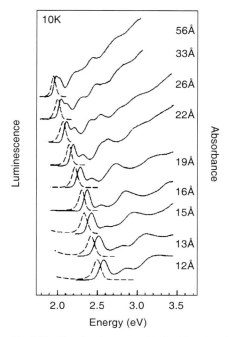

Fig. 2.17 Photoluminescence (dashed lines) and optical absorption (solid lines) of CdSe nanocrystal quantum dots for various nanocrystal radii. Reprinted with permission from [113]. Copyright (1996) by the American Physical Society.

picoseconds), and any multiexciton states typically have very short lifetimes, on the order of 100 ps. This short lifetime is due to Auger processes that allow an electron and hole to recombine, giving their energy to another electron–hole pair. Of course, there are various exceptions, where the higher energy states come into play. For example, measurements have been carried out on electrochemically doped nanocrystals [115], with electrons filling the lowest levels. Also, multi-exciton states can be observed, albeit on short timescales, which has significant interest for making efficient photovoltaic devices [116]. Nonetheless, in most cases the nanocrystal has at most one electron and/or hole in the lowest energy states.

When an electron and hole are excited in an NCQD, their initial spin polarization is determined by the selection rules discussed above. However, once they relax into the dark exciton ground state they remain there for so long that the spin polarization is often lost before recombination occurs. For this reason, the polarization of PL is not a good way to measure the spin in a NCQD. Instead, Faraday rotation has been used to measure the spin of electrons in such QDs, which will be discussed in detail in Chapter 7.

3
Theory of Confined States in Quantum Dots

In this chapter we discuss single-particle models to describe conduction-band electron and valence-band hole states in quantum dots. We start with a brief review of semiconductor physics in Section 3.1. We provide a description of quantum confinement and present simple models to describe the states of spherical and flat quantum dots that have proven useful for the theoretical description of structures introduced in Chapter 2. We conclude the chapter with a brief discussion of extensions of the single-particle model for realistic quantum dots, the Coulomb interaction, and an introduction to the concept of an exciton, a bound pair of an electron and a hole in quantum dots.

3.1
Band Structure of III–V Semiconductors

This section serves as the basis for our discussion of quantum dot states. While we assume that the reader is familiar with the basic concepts of solid state theory [117–119], we summarize here a few results that are useful for the description of semiconductor quantum dot states. We focus this theoretical discussion on III–V semiconductor compounds with zincblende structure, such as gallium arsenide (GaAs) or indium arsenide (InAs). Many current investigations are dedicated to quantum dots that are formed with such materials, for instance, self-assembled quantum dots. The III–V band structure fortunately also serves for the discussion of confined states of a second type of quantum dots, namely, small spherical quantum dot structures consisting of a II–VI semiconductor compound (such as cadmium selenide) with hexagonal crystal structure and an axial anisotropy [120].

The electronic band structure of a bulk semiconductor with zincblende structure is illustrated in Figure 3.1. The bands are parabolic close to their extrema, which are all located at the Γ point ($k = 0$), that is, the center of the Wigner–Seitz cell. The bottom of the conduction (c) band and the top of the valence (v) band are split by the bandgap energy E_g. The Fermi energy of an intrinsic (that is, undoped) semiconductor lies by definition within the bandgap. So in the crystal ground state, the v band is completely filled, while the c band is empty. The absence of an inversion center in the zincblende crystal leads to small shifts of the bands in k space [121].

Spins in Optically Active Quantum Dots. Concepts and Methods.
Oliver Gywat, Hubert J. Krenner, and Jesse Berezovsky
Copyright © 2010 WILEY-VCH Verlag GmbH & Co. KGaA, Weinheim
ISBN: 978-3-527-40806-1

For example, those terms induce a k dependent spin splitting of the c band. Regarding the energy splitting of confined levels in quantum dots, these energy shifts are small and we neglect them here.

3.1.1
Effective Mass of Crystal Electrons

There is an approximately parabolic energy dispersion around the extrema of the bands, as shown in Figure 3.1. The electrons thus behave like free effective particles, which are called crystal electrons [118]. In an isotropic and parabolic band b, the kinetic energy of a crystal electron can be written as $\hbar^2 k^2 / 2m_b^*$, where m_b^* is called the effective mass of the crystal electron. Generalizing this picture to anisotropic energy dispersions, the inverse effective mass is defined as a second-rank tensor that is obtained from the curvature of the energy dispersion $E_b(\mathbf{k})$,

$$\left[(m_b^*)^{-1}\right]_{\alpha\beta} = \frac{1}{\hbar^2}\frac{\partial^2 E_b(\mathbf{k})}{\partial k_\alpha \partial k_\beta}.\tag{3.1}$$

The effective mass of crystal electrons can be measured experimentally by cyclotron resonance [117]. In general, we apply the effective mass approximation whenever the intra-band dynamics of electrons is considered. When studying transitions between different bands, as in Section 5.2.2, it should be kept in mind that an

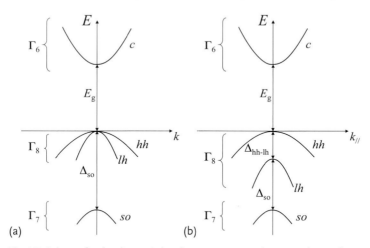

Fig. 3.1 Scheme for the electronic bandstructure in the vicinity of the Γ point for a three-dimensional crystal with zincblende lattice structure (a) without strain (b) in the presence of uniaxial strain. The bands shown are the conduction (c) band, the heavy-hole (hh) band, the light-hole (lh) band, and the spin–orbit split-off (so) band. Energies are given on an arbitrary scale as a function of (a) the wavevector k (b) the in-plane wavevector $k_\parallel$ (perpendicular to the strain axis). The band ordering displayed in (b) serves for the discussion of confined quantum dot states, where, in contrast to (a), the hh and the lh bands are split by Δ_{hh-lh}.

effective mass cannot be taken into account in a meaningful way and the picture of effective crystal electrons needs to be abandoned.

3.1.2
Spin–Orbit Interaction

The spin–orbit interaction is essential for the structure of the bands of Figure 3.1. At the heart of the spin–orbit interaction lies the relativistic effect that an electron moving in an external electric field experiences an effective magnetic field that couples to its spin $\mathbf{S}$. For an electron with momentum $\mathbf{p}$ in the presence of the electric potential Φ, an expansion in powers of $1/c$ of the Dirac equation yields in second-order the term

$$H_{so} = -\frac{e}{2\,m_0^2 c^2}\mathbf{S}\cdot(\nabla\Phi\times\mathbf{p})\,, \tag{3.2}$$

which is called the spin–orbit interaction term. We can see in the above expression that the effective magnetic field is generated by $\nabla\Phi\times\mathbf{p} = \mathbf{p}\times\mathbf{E}$, where $\mathbf{E}$ is the static electric field. If the spin–orbit interaction is nearly isotropic in the crystal, we obtain

$$H_{so} \approx \lambda_{so}\mathbf{L}\cdot\mathbf{S}\,, \tag{3.3}$$

where $\mathbf{L}$ is the orbital angular momentum of the electron in the semiconductor crystal and λ_{so} a constant. Then, with $\mathbf{J}^2 = (\mathbf{L}+\mathbf{S})^2$, the total angular momentum $\mathbf{J}$ of the electron provides a diagonal representation of H_{so}, and the corresponding quantum number j is a good quantum number in the semiconductor.

For an electron confined to a two-dimensional crystalline solid, the spin–orbit interaction Eq. (3.2) takes on a slightly more complicated form,

$$H_{so}^{2D} = \alpha_R(p_x\sigma_y - p_y\sigma_x) + \beta_D(-p_x\sigma_x + p_y\sigma_y) + O(p^3)\,, \tag{3.4}$$

where α_R is the Rashba coefficient, β_D the Dresselhaus coefficient, and $\mathbf{p} = (p_x, p_y)$ the two-dimensional momentum operator. The Rashba term [122] results from a structure inversion asymmetry of the confinement potential, whereas the Dresselhaus term [123] is due to the bulk inversion asymmetry of the crystal lattice. In a strictly two-dimensional system, the cubic Dresselhaus terms can be neglected relative to the linear terms, as given in Eq. (3.4).

3.1.3
Band Structure Close to the Band Edges

We now take a more detailed look at the band structure of a zincblende semiconductor in the region around the Γ point, as shown in Figure 3.1. In general, the c band and the v sub-bands, which together form the v band, are twofold degenerate if we neglect the small splittings due to the absence of inversion asymmetry, as mentioned above. At $k = 0$, the c band has total angular momentum $j = 1/2$. The

v band consists of three sub-bands, the heavy-hole (hh), the light-hole (lh), and the spin-orbit split-off (so) band.

By definition the v sub-bands are completely filled in the crystal ground state. As the most relevant dynamics of the v sub-bands is therefore due to missing electrons, called holes, the nomenclature refers to holes rather than electrons in some cases. At $k = 0$, the hh and the lh band have total angular momentum $j = 3/2$ and the so band has $j = 1/2$. The v band states with $j = 3/2$ and $j = 1/2$ are split by the spin–orbit interaction energy $\Delta_{so} = 3\hbar^2\lambda_{so}/2$, which is easily derived from Eq. (3.3). In the literature, the corresponding spinor representations are called Γ_6 for the c band, Γ_8 for the hh and the lh band, and Γ_7 for the so band [121, 124]. The splitting Δ_{so} is large in our cases of interest. For example, $\Delta_{so} = 340$ meV for GaAs and $\Delta_{so} = 380$ meV for InAs [125]. We therefore exclude the so states from our discussion in the following. The above classification of the $k = 0$ states follows directly from the symmetry properties of the semiconductor crystal and the atomic orbitals that give rise to the corresponding bands.

As can be seen in Figure 3.1, the hh and the lh sub-bands are degenerate at the Γ point and split into two branches for finite wave vectors k. The hh states have the angular momentum projections $J_z = \pm 3\hbar/2$ and the lh states $J_z = \pm\hbar/2$ at $k = 0$. The different curvatures of the two branches induce different effective masses for heavy and light holes according to Eq. (3.1); this is where their names come from. In Figure 3.1 we indicate parabolic bands, which is typically only applicable as an expansion to the regions close to $k = 0$. Usually, strong mixing occurs between the bands and sub-bands further away from the band extrema, especially between the relatively closely located hh and lh sub-bands. This mixing leads to a deviation from the parabolic dispersion relation. At large k, the total angular momentum j is thus no longer a good quantum number.

3.1.4
Band-Edge Bloch States

According to Bloch's theorem, electrons in a periodic crystal potential in the band b with wave vector $\mathbf{k}$ are described by a wave function

$$\langle\mathbf{r}|\Psi_{\mathbf{k}}^b\rangle = \mathrm{e}^{\mathrm{i}\mathbf{k}\cdot\mathbf{r}}u_{\mathbf{k}}^b(\mathbf{r})\,. \tag{3.5}$$

Here, we apply Dirac's bra-ket notation on the left-hand side. The Bloch state $u_{\mathbf{k}}^b(\mathbf{r})$ on the right-hand side is a function that is periodic in the Bravais lattice of the crystal. In the following we consider quantum dots that, due to their spatial confinement in three dimensions, provide fully quantized states for crystal electrons. We consider only values of k close to zero, such that it is a good approximation to classify the states by the total angular momentum j. It is actually not clear from the beginning that j is a good quantum number for confined quantum dot states, especially for small dots with radii of only a few nanometers. For confined states, decreasing spatial dimensions lead to an increase of the wave number k. This gives more weight to even higher-order terms in the energy dispersion, which enhances the coupling to bands with different angular momenta. However, experimental data has suggested so far that the small k assumption provides a reasonable approximation for the Bloch states for the types of quantum dots discussed in this book. Consequently, for the states confined in quantum dots we write for the Bloch states in the following $u_{J_z}^{(j)}(\mathbf{r})$, using the total angular momentum j and the projection J_z as labels instead of b and $\mathbf{k}$.

The Bloch states of the c band in a semiconductor quantum dot with zincblende structure have orbital s symmetry,

$$|u_{+1/2}^{(1/2)}\rangle = |s\rangle|\uparrow\rangle , \quad |u_{-1/2}^{(1/2)}\rangle = |s\rangle|\downarrow\rangle , \tag{3.6}$$

where $|s\rangle$ is a function that is invariant under all symmetry transformations of the Bravais lattice. We define the spin states $|\uparrow\rangle$ and $|\downarrow\rangle$ along a crystal main axis z, for example [001]. The above c band edge states are spin degenerate if we neglect small terms that are cubic in k and which are due to the absence of inversion symmetry in the zincblende lattice.

The v band Bloch states have orbital p symmetry. By coupling the orbital angular momentum $l = 1$ and the spin $S = 1/2$ according to the usual Clebsch–Gordan theory, the hh states are obtained as

$$|u_{+3/2}^{(3/2)}\rangle = -\frac{1}{\sqrt{2}}|x+iy\rangle|\uparrow\rangle , \tag{3.7}$$

$$|u_{-3/2}^{(3/2)}\rangle = \frac{1}{\sqrt{2}}|x-iy\rangle|\downarrow\rangle , \tag{3.8}$$

and the lh states as

$$|u_{+1/2}^{(3/2)}\rangle = -\frac{1}{\sqrt{6}}\left(|x+iy\rangle|\downarrow\rangle + 2|z\rangle|\uparrow\rangle\right), \tag{3.9}$$

$$|u_{-1/2}^{(3/2)}\rangle = \frac{1}{\sqrt{6}}\left(|x-iy\rangle|\uparrow\rangle + 2|z\rangle|\downarrow\rangle\right). \tag{3.10}$$

In the above expressions, $|x \pm iy\rangle = |x\rangle \pm i|y\rangle$ and the states $|a\rangle$ are the p-type parts of the Bloch states, which transform like the coordinates $a = x, y, z$. Again, we omit here the so states (with $j = 1/2$) because they are far away in energy.

Looking at the above hh and lh states again, we note that for the hh states there is a one-to-one correspondence between the projections of the total angular momentum and the spin. For lh states, however, both spin orientations are present in both angular momentum projections. As we discuss further in Section 5.2.2, this allows for more lh optical transitions than for hh states.

3.1.5
Coupling of Bands and the Luttinger Hamiltonian

For the discussion of the dynamics of electrons in bands around a specific point $\mathbf{k} = \mathbf{k}_0$ in k-space, the so-called $\mathbf{k}\cdot\mathbf{p}$ theory is a very powerful tool. We do not attempt a complete discussion of this topic here, and only summarize a few important results. For our purposes, $\mathbf{k} \cdot \mathbf{p}$ theory provides a description in perturbation theory of the band energies and states near the band extrema at $\mathbf{k}_0 = 0$. We stick with writing $\mathbf{k}_0$ for the expansion point below in favor of a more general notation. The band-edge Bloch states of the type as given in Eqs. (3.6)–(3.10) above serve as an orthogonal basis for the perturbation expansion in small k around $\mathbf{k}_0$. Obviously, the symmetry properties of the band-edge Bloch states determine the spectrum near the band extremum obtained in the frame of this theory. The expansion of the energy $E_n(\mathbf{k}_0)$ of the band $b = n$ in small wave vectors $\mathbf{k}$ around $\mathbf{k}_0$ is the first result of $\mathbf{k} \cdot \mathbf{p}$ theory that we show. In the discussion of the quantum dot states later in this chapter we will come back to this result. We assume that $E_n(\mathbf{k}_0)$ is an extremum of the band n, and that the corresponding electron state is f-fold degenerate with kets $|n_1\rangle, \ldots, |n_f\rangle$. To second-order in degenerate perturbation theory we obtain for the Hamiltonian

$$
\langle n_j | H | n_i \rangle = \delta_{ij} \left[E_n(\mathbf{k}_0) + \frac{\hbar^2 k^2}{2m_0} \right]
$$
$$
+ \frac{\hbar^2}{m_0^2} \sum_{m \neq n_s} \frac{\langle n_j | \mathbf{k} \cdot \mathbf{p} | m \rangle \langle m | \mathbf{k} \cdot \mathbf{p} | n_i \rangle}{E_n^{(0)} - E_m^{(0)}} , \tag{3.11}
$$

where $E_n^{(0)}$ denotes the zeroth-order energy of the unperturbed states $|n_s\rangle$, with the index $s = 1, \ldots, f$ for the degeneracy. In the sum over all m we can usually exclude $m = n_s$ for all s because the momentum matrix element vanishes if taken between two states with the same parity. The second-order correction with the characteristic $\mathbf{k} \cdot \mathbf{p}$ matrix elements has given the name to the theory. In those matrix elements, $\mathbf{k}$ serves as the expansion parameter in the sense of perturbation theory, while $\mathbf{p}$ is an operator acting on the states of the matrix element. To indicate this, many authors take $\mathbf{k}$ in front of the momentum matrix elements. In principle, $\mathbf{k} \cdot \mathbf{p}$ theory can be expanded in an arbitrary number of bands.

Another important result from multiband $\mathbf{k} \cdot \mathbf{p}$ theory is the expression for the effective gyromagnetic (g) factor of electrons in the c band with energy E (taken with respect to the band minimum energy, $E = 0$), [126, 127]

$$
g_c(E) \approx g_0 - \frac{2}{3} \frac{E_p \Delta_{so}}{(E_g + E)(E_g + \Delta_{so} + E)} . \tag{3.12}
$$

Here, $g_0 = 2.00$ is the free electron g factor and $E_p = 2(\langle s | \mathbf{p} | x \rangle)^2 / m_0$ is the so-called Kane energy with the momentum matrix element taken between an s-like and a p-like Bloch state.

If we focus on two valence sub-bands only and write down the Hamiltonian with the four $j = 3/2$ Bloch states Eqs. (3.7)–(3.10) for the states $|n_s\rangle$ and set $E_n(\mathbf{k}_0) = 0$, then we obtain the famous Luttinger Hamiltonian [128],

$$
H_L = \begin{pmatrix} H_{hh} & c & b & 0 \\ c^* & H_{lh} & 0 & -b \\ b^* & 0 & H_{lh} & -c \\ 0 & -b^* & -c^* & H_{hh} \end{pmatrix}, \tag{3.13}
$$

where the asterisk * denotes complex conjugation, and we use the following abbreviations,

$$H_{hh} = \frac{\hbar^2}{2m_0}(\gamma_1 + \gamma_2)\left(k_x^2 + k_y^2\right) + \frac{\hbar^2}{2m_0}(\gamma_1 - 2\gamma_2)k_z^2 \tag{3.14}$$

$$H_{lh} = \frac{\hbar^2}{2m_0}(\gamma_1 - \gamma_2)\left(k_x^2 + k_y^2\right) + \frac{\hbar^2}{2m_0}(\gamma_1 + 2\gamma_2)k_z^2 \tag{3.15}$$

$$c = \frac{\hbar^2\sqrt{3}}{m_0}\gamma_3 k_z(k_x - ik_y) \tag{3.16}$$

$$b = \frac{\hbar^2\sqrt{3}}{2m_0}\gamma_2\left(k_x^2 - k_y^2\right) - i\frac{\hbar^2\sqrt{3}}{m_0}\gamma_3 k_x k_y \ . \tag{3.17}$$

In above equations, γ_1, γ_2, and γ_3 are the Luttinger parameters. The values of the Luttinger parameters γ_i depend on the strength of the inter-band coupling and are obtained experimentally. They determine the effective masses of hh and lh states as indicated in H_{hh} and H_{lh}. It is obvious from these expressions that the effective masses of the hh and the lh bands are usually anisotropic. Sometimes in the literature a different representation is chosen for the $j = 3/2$ angular momentum eigenfunctions than Eqs. (3.7)–(3.10), for example the Luttinger–Kohn representation [129]. This leads to different phases of some matrix elements in H_L compared to the standard representation, but of course the energy eigenvalues are the same. The Luttinger Hamiltonian H_L can be written down in a more compact form by using the $j = 3/2$ angular momentum matrices,

$$J_x = \frac{\hbar}{2}\begin{pmatrix} 0 & \sqrt{3} & 0 & 0 \\ \sqrt{3} & 0 & 2 & 0 \\ 0 & 2 & 0 & \sqrt{3} \\ 0 & 0 & \sqrt{3} & 0 \end{pmatrix}, \tag{3.18}$$

$$J_y = \frac{i\hbar}{2}\begin{pmatrix} 0 & -\sqrt{3} & 0 & 0 \\ \sqrt{3} & 0 & -2 & 0 \\ 0 & 2 & 0 & -\sqrt{3} \\ 0 & 0 & \sqrt{3} & 0 \end{pmatrix}, \tag{3.19}$$

$$J_z = \frac{\hbar}{2}\begin{pmatrix} 3 & 0 & 0 & 0 \\ 0 & 1 & 0 & 0 \\ 0 & 0 & -1 & 0 \\ 0 & 0 & 0 & -3 \end{pmatrix}. \tag{3.20}$$

Using above matrices J_α and the anticommutator $\{J_\alpha, J_\beta\} = J_\alpha J_\beta + J_\beta J_\alpha$, we can alternatively write for the Luttinger Hamiltonian

$$H_L = -\frac{\hbar^2}{2m_0}\left(\gamma_1 + \frac{5}{2}\gamma_2\right)k^2 + \frac{\hbar^2}{m_0}\gamma_2\left(k_x^2 J_x^2 + k_y^2 J_y^2 + k_z^2 J_z^2\right)$$
$$+ \frac{\hbar^2}{m_0}\gamma_3\left(k_x k_y\{J_x, J_y\} + k_x k_z\{J_x, J_z\} + k_y k_z\{J_y, J_z\}\right). \tag{3.21}$$

The Luttinger Hamiltonian is sometimes also shown in a form using the momentum operator, which is obtained from above expressions by replacing $k_\alpha \to p_\alpha/\hbar$.

There are two frequently encountered approximations for the Luttinger parameters. In the so-called spherical approximation the parameters γ_2 and γ_3 are replaced in H_L by the value

$$\gamma = \frac{1}{5}(2\gamma_2 + 3\gamma_3) . \tag{3.22}$$

As a consequence of this substitution, the warping of the valence bands due to the cubic symmetry is neglected. The Luttinger Hamiltonian Eq. (3.21) can then be written in the compact form

$$H_{L,\text{spherical}} = -\frac{\hbar^2}{2 m_0} \left(\gamma_1 + \frac{5}{2}\gamma \right) k^2 + \frac{\hbar^2}{m_0} \gamma \, (\mathbf{k} \cdot \mathbf{J})^2 . \tag{3.23}$$

In our study of the confined quantum dot states below, we will come back to this form of the ν band Hamiltonian.

Another approximation for the Luttinger parameters is the so-called axial approximation. We do not apply this more special approximation in this book, but mention it for reasons of completeness. In the axial approximation, γ_2 and γ_3 are replaced in the b matrix elements by their average

$$\gamma' = \frac{1}{2}(\gamma_2 + \gamma_3) . \tag{3.24}$$

After this modification, H_L is symmetric about the z axis, and the warping of the valence bands is neglected in the xy plane only.

3.1.6
Splitting of Heavy Hole and Light Hole Bands

Uniaxial strain in the semiconductor crystal is a mechanism that can lift the degeneracy of heavy and light holes at $k = 0$ [121]. For uniaxial compression in GaAs, the top of the lh band is higher in energy than the top of the hh band. For uniaxial stretch, the opposite ordering is achieved. We assume now that the hh and lh bands are split at $k = 0$ by an energy Δ_{hh-lh}. If the deformation axis z' does not coincide with z, a new angular momentum quantization axis is defined for the ν band states. The energy splitting Δ_{hh-lh} quenches the transition probabilities between heavy and light hole states, which is immediately clear in the expression for transition rates in perturbation theory, where the energy difference enters in the denominator. According to Chapter 5, the dominant transition mechanism among the levels of the $j = 3/2$ manifold are single-photon magnetic dipole transitions with a total angular momentum difference $\Delta J_z = \pm\hbar$. Hence, the dominant hole angular momentum transitions are ordered in the sequence

$$J_z = +\frac{3\hbar}{2} \leftrightarrow J_z = +\frac{\hbar}{2} \leftrightarrow J_z = -\frac{\hbar}{2} \leftrightarrow J_z = -\frac{3\hbar}{2} . \tag{3.25}$$

The direct transition between the two outermost hh states in above sequence obviously requires an exchange of angular momentum $\Delta J_z = \pm 3\hbar$. The leading terms of such a transition are due to higher-order processes such as, for example, a three-photon transition or a combination of photon and phonon transitions, which usually have a very low transition probability. We conclude from these considerations that with increasing splitting Δ_{hh-lh}, the hh angular momentum states become "frozen" under usual conditions, when compared to the bulk where the hole angular momentum relaxes typically faster than picoseconds. An important mechanism that can flip a hole spin is given by the electron–hole exchange interaction, see Section 6.2. However, to become active, the electron–hole exchange interaction requires the presence of an electron in addition to the hole, of course.

3.1.7
Electrons and Holes

Via an external excitation of the semiconductor, for example by the absorption of a photon, an electron in a v band state can be excited into a c band state. This process leaves an unoccupied state in the v band, which is called a hole. As unoccupied states give rise to the intraband dynamics in the otherwise filled v band, it is usually more intuitive to consider holes in these cases, for example for the confinement in a quantum dot structure. In contrast, we usually stick with the electron picture when calculating transition matrix elements, for simplicity. In Chapter 5 we investigate in more depth the optical selection rules that need to be satisfied for inter-band transitions related to radiation.

In the hole picture we take into account that there is a net positive charge associated with an empty valence band state. This is the result of the positive nuclear charge, which is not neutralized by the surrounding electrons. As band states are delocalized, the unoccupied state moves through the crystal and effectively behaves like a particle with a positive charge. We thus may consider the empty v band state as an effective particle with positive elementary charge e and the effective mass of the corresponding band.

3.2
Quantum Confinement

For a single particle with momentum p, quantum mechanical behavior generally becomes apparent when it moves in a potential that varies significantly on a length scale comparable to its de Broglie wavelength $\lambda_p = h/p$, where h is Planck's constant. As we already mentioned earlier in this book, semiconductor quantum dots with an extension in all three dimensions on the order of nanometers to tens of nanometers exhibit a discrete energy spectrum of fully quantized electron states. Here we discuss a few important concepts and models of quantum confinement.

3.2.1
One-Dimensional Confinement

We start our discussion of confined states with a brief example, the quantum confinement along one crystal axis z due to a deep rectangular potential well. This model is usually applied to quantum wells with growth direction z, but is also relevant to quantum dots with a flat, that is, oblate, shape. The confinement along z quantizes the z-component p_z of the momentum operator. Already at this level we can see that the hh and lh states with a given p_z are split at the Γ point ($k = 0$) by a confinement energy

$$|\Delta^z_{hh-lh}| = \langle p_z^2 \rangle \left(\frac{1}{2m_{lh}} - \frac{1}{2m_{hh}} \right), \tag{3.26}$$

with the effective masses taken along z, and the usual quantum mechanical expectation value, $\langle \ldots \rangle$. Obviously, one-dimensional confinement raises the lh band above the hh band in energy. Because of the lifting of the hh–lh degeneracy along the crystal axis z, the confinement asymmetry defines an angular momentum quantization axis for the v band states along z. In practice, the presence of strain also needs to be taken into account for the v band structure, such as the uniaxial deformations discussed in Section 3.1.6. Actually, in self-assembled quantum dots, the strain overcompensates the sub-band reordering due to the asymmetric confinement in many cases and leads to a resulting band ordering as illustrated in Figure 3.1b.

3.2.2
Quantum Dot Confinement

For an electron in the band b that is confined to a quantum dot, the equation of motion takes on quite a simple form in the effective mass approximation. It becomes the motion of a crystal electron with an effective mass m_b^* in an effective potential $V_{qd,b}(\mathbf{r})$ that characterizes the shape of the quantum dot. These effective quantities spare us the more difficult treatment of the Schrödinger equation in presence of the full crystal potential of the quantum dot. The effective potentials $V_{qd,b}(\mathbf{r})$ may be different for different bands b, since they are valid for crystal electrons in different branches of the energy dispersion relation.

Another approximation that we apply to quantum dot states in this book is the envelope function approximation. In this approximation, the total wave function $|\Psi_b\rangle$ of a single electron in the band b is written as

$$\langle \mathbf{r} | \Psi_b \rangle = \sum_{l,m;j,J_z} c_{l,m}^{j,J_z} \, \psi_{lm}(\mathbf{r}) \, u_{J_z}^{(j)}(\mathbf{r}) \,, \tag{3.27}$$

with the Bloch functions $u_{J_z}^{(j)}(\mathbf{r})$, envelope functions $\psi_{lm}(\mathbf{r})$, and expansion coefficients $c_{l,m}^{j,J_z}$. This wave function is similar to the wave function Eq. (3.5) for Bloch electrons. However, in contrast to the plane wave function $\exp(i\mathbf{k} \cdot \mathbf{r})$ of Bloch electrons in the crystal, the envelope functions $\psi_{lm}(\mathbf{r})$ are localized at the quantum dot according to the effective quantum dot potentials $V_{qd,b}(\mathbf{r})$ of the specific band b. The indices l, m are the angular momentum quantum numbers for spherical quantum dots. For other types of quantum dots, a different basis may be more suitable. The envelope function approximation Eq. (3.27) can be justified only for quantum dots for which the potential $V_{qd,b}(\mathbf{r})$ varies over a length scale much larger than the lattice constant.

Referring back to our previous discussion of the splitting of heavy and light hole states, we note that typical $hh-lh$ splittings found in quantum dot experiments are on the order of $\Delta_{hh-lh} \sim 10\,\text{meV}$ across many different systems. Usually, the energetically lowest interband transitions between the c and v bands in quantum dots are hh transitions, indicating that the v band ordering is mainly determined by strain.

In the following, we discuss two examples of experimentally relevant quantum dot models. The first model is for spherical quantum dots, which are usually chemically synthesized colloidal dots, and the second model is for parabolic, lens-shaped self-assembled quantum dots.

3.3
Spherical Quantum Dot Confinement

Spherically symmetric quantum dot structures are special due to their maximum spatial symmetry. As is well known from quantum mechanics, the Schrödinger equation with a spherically symmetric potential can be reduced to a one-dimensional problem, namely, the radial Schrödinger equation. For the most simple cases, this allows for an analytical solution of the confined states in a spherical quantum dot structure. Spherical quantum dots are usually surrounded by another material into which the confined wave functions may protrude. It is sometimes desirable to cover a quantum dot with a large-bandgap material, for example to avoid photoluminescence bleaching due to carrier trapping in surface states. Even more elaborate systems are nested spherical shells that surround a core, separated by interlying large-bandgap barriers.

We discuss here a generic system for the above mentioned structure types, namely, a spherical quantum dot (which we refer to as the core) embedded in an infinitely large shell that consists of a different material than the core. The interface between the core and the shell deserves special consideration due to the abrupt jump in the effective mass of the band electrons. To solve this problem, an ansatz for the lowest-energy wave function pieces is made and then the conditions are shown that need to be satisfied to connect a wave function piece in the core with a piece in the shell.

3.3.1
Conduction-Band States

We consider a core made of a semiconductor a, with radius r_0, surrounded by an infinitely large shell made of a semiconductor b. Before we solve the Schrödinger equation for a c band electron in this core-shell structure, we note that its effective mass is position dependent,

$$m_c^*(r) = m_a \quad \text{for} \quad r < r_0 , \quad \text{and} \quad m_c^*(r) = m_b \quad \text{for} \quad r > r_0 . \tag{3.28}$$

In this situation, the usual kinetic energy operator $\mathbf{p}^2/2m_c^*(r)$ (or, alternatively, also $[1/2m_c^*(r)]\mathbf{p}^2$) is not Hermitian because the radial part of the momentum operator and a function of r do not commute. However, we may obtain a Hermitian form of the kinetic energy by an ordering of operators such as

$$H_{0,c} = \mathbf{p}\frac{1}{2m_c^*(r)}\mathbf{p} . \tag{3.29}$$

The c band Hamiltonian for the envelope function is in the two-band approximation and in effective-mass theory given by

$$H = H_{0,c} + V_c(r) , \tag{3.30}$$

where $H_{0,c}$ is in a Hermitian form as given above and $V_c(r)$ is the effective radial potential of the core-shell structure. We set the c band energy of the core to zero and assume a step-like radial potential,

$$V_c(r) = V_c \Theta(r - r_0) , \tag{3.31}$$

where $\Theta(r)$ is the Heaviside step function. We now consider $r \neq r_0$ and discuss the solutions away from the boundary between the core and the shell.

In envelope function theory, Eq. (3.27), the electron wave function can be represented as

$$\Psi_c(\mathbf{r}) = \sum_{l,m;j,J_z} c_{l,m}^{j,J_z} Y_{lm}(\hat{\mathbf{r}}) R_l(r) u_{J_z}^{(j)}(\mathbf{r}) , \tag{3.32}$$

where the envelope function has the well known structure with the angular momentum eigenfunctions Y_{lm} and the radial eigenfunctions R_l, and $c_{l,m}^{j,J_z}$ are the Clebsch–Gordan coefficients. Further, $\hat{\mathbf{r}} = \mathbf{r}/r$ is a radial unit vector. It can be seen in Eq. (3.32) that for spherical quantum dots there are two important kinds of angular momenta: There is the total angular momentum $\mathbf{J}$ of the Bloch function, which is the angular momentum of the band, and there is the orbital angular momentum $\mathbf{L}$ of the envelope function $Y_{lm}(\hat{\mathbf{r}})R_l(r)$. Electron states in a spherically symmetric structure can be classified according to the total angular momentum $\mathbf{F} = \mathbf{L} + \mathbf{J}$ and the parity operator $\mathbf{P}$, which acts on the wave function as $\mathbf{P}\Psi(\mathbf{r}) = \Psi(-\mathbf{r})$. Note that $\mathbf{J}$ is invariant under the action of $\mathbf{P}$ because it is a band angular momentum.

For the c band states the internal angular momentum is $j = 1/2$, and therefore each of the eigenstates $|f, F_z\rangle$ is composed of two parts with orbital angular momenta $l = f \pm \hbar/2$, respectively. Yet, these two parts have different parities because of their orbital components, as $\mathbf{P} Y_{lm} = (-1)^l Y_{lm}$. A classification according to $\mathbf{F}$, $\mathbf{P}$, and n, where n enumerates the radial solutions for a given l, is therefore equivalent to a classification according to the quantum numbers (n, l, m, J_z) for the conduction band states. We write for the conduction-band (electron) eigenstates

$$\langle \mathbf{r} | n l_m J_z \rangle = Y_{lm}(\hat{\mathbf{r}}) R_l(r) u_{J_z}^{(1/2)}(\mathbf{r}) , \tag{3.33}$$

where R_l satisfies at $r \neq r_0$ the radial Schrödinger equation with angular momentum quantum number l,

$$\left[-\partial_r^2 - \frac{2}{r} \partial_r + \frac{l(l+1)}{r^2} - k_{nl}^2 \right] R_l(r) = 0 . \tag{3.34}$$

Here, ∂_r is the partial derivative with respect to the radial coordinate r. In terms of the eigenenergies E_{nl} we have defined in the above equation

$$k_{nl}^2 = \frac{2 m_c^*(r)}{\hbar^2} \left[E_{nl} - V_c(r) \right] . \tag{3.35}$$

In the following, we consider the case of zero external magnetic field and drop the spin index J_z. As V_c is constant for $r \neq r_0$, the radial solution of Eq. (3.34) is given there by

$$R_l(r) = c_1 j_l(k_{nl} r) + c_2 y_l(k_{nl} r) , \tag{3.36}$$

with parameters $c_{1,2}$ and j_l and y_l are the l-th order spherical Bessel function of first and second kind, respectively. In terms of the Bessel functions J_ν and Y_ν, the spherical Bessel functions are given by $j_l(x) = \sqrt{\pi/2x}\, J_{l+1/2}(x)$ and $y_l(x) = \sqrt{\pi/2x}\, Y_{l+1/2}(x)$.

For $l = 0$, the spherical Bessel functions of first and second kind read

$$j_0(r) = \frac{\sin r}{r} , \quad y_0(r) = -\frac{\cos r}{r} . \tag{3.37}$$

For states in the classically forbidden region, a suitable basis is given by the spherical Hankel functions of first and second kind, $h_l^{\pm}(ir) = j_l(ir) \pm i y_l(ir)$, of which the zeroth-order states are given by

$$h_0^+(ir) = -\frac{e^{-r}}{r} , \quad h_0^-(ir) = \frac{e^r}{r} . \tag{3.38}$$

Since $y_l(r)$ diverges at $r = 0$ for all l, the wavefunction in the core is a spherical Bessel function of the first kind. In order to connect the solutions of the wave functions for the core and the shell piece by piece, we write for the ansatz in the core the wave function

$$R_l^{\text{core}}(r) = a_1 \frac{j_l(k_{nl,\text{core}} r)}{j_l(k_{nl,\text{core}} r_0)} , \tag{3.39}$$

where a normalization and the parameter a_1 have been introduced for convenience. For the classically forbidden wavefunction in the shell, the radial wave vector is imaginary, $k_{nl,\text{shell}} = i\kappa_{nl,\text{shell}}$, and we choose the ansatz

$$R_l^{\text{shell}}(r) = a_2 \frac{h_l^{(+)}(i\kappa_{nl,\text{shell}}r)}{h_l^{(+)}(i\kappa_{nl,\text{shell}}r_0)} + a_3 \frac{h_l^{(-)}(i\kappa_{nl,\text{shell}}r)}{h_l^{(-)}(i\kappa_{nl,\text{shell}}r_0)} . \tag{3.40}$$

For the classically allowed shell radial states, that is, the states with an energy larger than V_c, a suitable ansatz is given by

$$R_l^{\text{shell}}(r) = a_2 \frac{j_l(k_{nl,\text{shell}}r)}{j_l(k_{nl,\text{shell}}r_0)} + a_3 \frac{y_l(k_{nl,\text{shell}}r)}{y_l(k_{nl,\text{shell}}r_0)} . \tag{3.41}$$

The total radial wavefunction is assembled as follows,

$$R_l(r) = \frac{1}{\sqrt{N}} \left[R_l^{\text{core}}(r)\Theta(r_0 - r) + R_l^{\text{shell}}(r)\Theta(r - r_0) \right], \tag{3.42}$$

where N is a normalization constant. The shell wave function ansatz is chosen according to whether the state under investigation is confined by the shell or propagates through it at an energy above V_c. The coefficients a_1, a_2, and a_3 contained in the above equation are determined by the boundary conditions of the Schrödinger equation. The first set of boundary conditions to be satisfied at $r = r_0$ is obtained from postulating the continuity of the wavefunction. If we assume that the Bloch states of the two crystals in the core and the shell are approximately equal and nonzero at the interface $r = r_0$, then we see that the continuity of the envelope functions is necessary at $r = r_0$. For the second set of conditions to connect the core and the shell states, we note that the frequently used assumption of the continuity of the derivative of the wave functions at the boundaries cannot be made if the effective mass is position dependent. Instead, we need to go back to the underlying principle of the continuity of the probability current [130]. To obtain the condition for the probability current, we multiply Eq. (3.34) from the left-hand side with r^2 and obtain

$$-\partial_r \left(r^2 \partial_r R_l \right) + \left[l(l + 1) - r^2 k_{nl}^2 \right] R_l = 0 . \tag{3.43}$$

We integrate this equation radially over the boundary r_0 in the region $r_0 - \delta \ldots r_0 + \delta$, which yields for $\delta \to 0$ the condition

$$\frac{1}{m_{\text{core}}^*} \partial_r R_l^{\text{core}}(r_0) = \frac{1}{m_{\text{shell}}^*} \partial_r R_l^{\text{shell}}(r_0) \tag{3.44}$$

for the continuity of the probability current. In the above equation, the effective masses have been taken corresponding to the region from which the boundary is approached. For $m_{\text{core}}^* \neq m_{\text{shell}}^*$, the above condition Eq. (3.44) induces a kink in the envelope function at the interface, in clear disagreement with the simplified assumption of the continuity of the derivative of the wavefunction.

The above boundary conditions have been found to give a reasonably good approximation for the *c* band states of spherically symmetric heterostructures, which we discussed in the beginning of Section 3.3. An alternative approach to the treatment of abrupt interfaces is to redefine the Bloch states for the entire heterostructure [130].

We conclude our discussion of the *c* band states with a few general remarks. In the derivation of the *c* band states above we have applied a number of strong assumptions and drastic simplifications. It is clear that the model discussed above is phenomenological and has strict limitations. It can be expected that ab-initio calculations that take into account the full crystal potential of the quantum dots provide a more accurate description of the confined states, though at the price of larger computational complexity. Quite surprisingly, the model discussed above (in combination with the *v* band model discussed below), which is based on the effective mass approximation and the envelope function approximation, has been shown to provide a good description of the observed spectra not only for small spherical quantum dots, but also for core-shell structures such as a single shell (sometimes called quantum-dot quantum-well) surrounded by spherical barriers made of a different material [131], and structures consisting of a core coupled to a shell through an interlying barrier [132]. While the observed good agreement might be coincidental, future investigations will show whether these models also work well for other structures with different sizes and materials.

3.3.2
Valence Band States

For the confined *v* band states we take into account the *hh* and the *lh* band. According to **k · p**-theory, the Hamiltonian for the *v* band is at $r \neq r_0$ in the four-band approximation and in effective-mass theory given by

$$
H = H_{0,v} + V_v(r)
$$

$$
= \left[\gamma_1(r) + \frac{5}{2}\gamma(r) \right] \frac{\mathbf{p}^2}{2m_0} - \frac{\gamma(r)}{m_0} (\mathbf{p} \cdot \mathbf{J})^2 + V_v(r) , \tag{3.45}
$$

where $H_{0,v}$ is the Luttinger Hamiltonian in the spherical approximation, see Eq. (3.23). The vector of angular momentum 3/2 operators is denoted by **J**, and the Luttinger parameters $\gamma_1(r)$ and $\gamma(r)$ are defined as

$$
\gamma_1(r) = \begin{cases} \gamma_{1,\text{core}} & \text{for} \quad r < r_0 \\ \gamma_{1,\text{shell}} & \text{for} \quad r_0 < r , \end{cases} \tag{3.46}
$$

and

$$
\gamma(r) = \begin{cases} \gamma_{\text{core}} & \text{for} \quad r < r_0 \\ \gamma_{\text{shell}} & \text{for} \quad r_0 < r . \end{cases} \tag{3.47}
$$

We apply the usual basis of the band-edge Bloch states u_{J_z} with angular momentum $j = 3/2$ in the following. The radial potential $V_v(r)$ for the *v* band is defined similarly to $V_c(r)$ as a spherical barrier potential. The v band states can be represented in envelope function theory according to Eq. (3.32), where we insert the $j = 3/2$ manifold. In contrast to the c band states, the classification of the v band states according to the total angular momentum **F** and the parity operator **P** provides coupling of the radial states l and $l + 2$ in the envelope functions, leading to $s - d$ coupling, $p - f$ coupling, and so on. We represent the v band states as $|nl_f; F_z\rangle$, which are obtained by applying the usual relations of Clebsch–Gordan coefficients. See Appendix A for the $f = 3/2$ and $f = 1/2$ states with even and odd parity. Obviously, the mixing of hh and lh states, which occurs along with the coupling of l and $l + 2$ states, leads to more complicated expressions for the v band states than for the c band states.

The solutions for the radial envelope wavefunctions R_l and R_{l+2} are obtained from the following set of coupled differential equations [133, 134],

$$\begin{pmatrix} A & B \\ C & D \end{pmatrix} \begin{pmatrix} R_l \\ R_{l+2} \end{pmatrix} = E \begin{pmatrix} R_l \\ R_{l+2} \end{pmatrix}, \tag{3.48}$$

where we have introduced the differential operators

$$A = -\frac{\hbar^2}{2m_0}(\gamma_1 + c_1\gamma)\left(\partial_r^2 + \frac{2}{r}\partial_r - \frac{l(l+1)}{r^2}\right) + V_v(r), \tag{3.49}$$

$$B = c_2\gamma\frac{\hbar^2}{2m_0}\left(\partial_r^2 + \frac{2l+5}{r}\partial_r + \frac{(l+1)(l+3)}{r^2}\right), \tag{3.50}$$

$$C = c_2\gamma\frac{\hbar^2}{2m_0}\left(\partial_r^2 - \frac{2l+1}{r}\partial_r + \frac{l(l+2)}{r^2}\right), \tag{3.51}$$

$$D = -\frac{\hbar^2}{2m_0}(\gamma_1 - c_1\gamma)\left(\partial_r^2 + \frac{2}{r}\partial_r - \frac{(l+2)(l+3)}{r^2}\right) + V_v(r). \tag{3.52}$$

Equation (3.48) is solved in terms of the spherical Bessel functions of first and second kind (or a linear superposition of them, such as the spherical Hankel functions), similarly as for the c band.

Connecting the core and shell wave function pieces for the v band at the interface is achieved by applying the same postulates as for the c band wave functions, namely, the continuity of the wave function and the continuity of the probability current. The latter condition requires the continuity of the expression

$$\begin{pmatrix} \gamma_1\partial_r + c_1\gamma\left(\partial_r + \frac{3}{2r}\right) & -c_2\gamma\left(\partial_r + \frac{l+3}{r}\right) \\ -c_2\gamma\left(\partial_r - \frac{l}{r}\right) & \gamma_1\partial_r - c_1\gamma\left(\partial_r + \frac{3}{2r}\right) \end{pmatrix} \begin{pmatrix} R_l \\ R_{l+2} \end{pmatrix} \tag{3.53}$$

at the boundary $r = r_0$.

In many experiments, cadmium selenide (CdSe) is used as the material of choice for spherical quantum dots. For CdSe with wurtzite crystal structure, there is a lattice anisotropy present that splits hh and lh states. This can be taken into account by the anisotropy Hamiltonian [120]

$$H_{\text{an}} = \Delta \left[\left(\frac{3}{2} \right)^2 - J_z^2 \right].$$

(3.54)

For example, for bulk CdSe, $\Delta = 25$ meV [120, 135]. This lifts the degeneracy of an l_f-multiplet.

The ansatz described above can be readily extended to nested spherical heterostructures containing more than just one interface between two different materials. For example, for coupled core-shell structures, this simple model has provided a surprisingly accurate description of the observed photoluminescence spectra [132]. While the envelope function approximation and also the effective mass approximation can, strictly speaking, not be applied to structures extending only over a few monolayers, the above model has been applied as an asymptotic guess, and surprising coincidence of energies has been obtained for the core-shell structures under study.

3.3.3
Deviations from a Spherical Dot Shape

Meier and Awschalom have shown that a broken spherical symmetry can account for a mixing of the $1S_{3/2}$ and $1P_{3/2}$ valence band multiplets [134], which was visible in experimental Faraday rotation data [131]. To take this deformation into account in a simple model, an additional potential term may be included in the Hamiltonian,

$$\delta V(\mathbf{r}) = v_0 \sin \theta \, (1 + \cos \phi) \,,$$

(3.55)

which mixes, for example, the states $1S_{3/2}$ and $1P_{3/2}$. The angles θ and ϕ denote the azimuthal and polar angle of $\mathbf{r}$ relative to the lattice symmetry axis, respectively, and v_0 is a fit parameter. The admixture of s-type to p-type multiplets redistributes the spectral weight and increases the number of resonances with comparable amplitude in the Faraday rotation signal. Further, the redistribution of spectral weight also explains the absence of pronounced resonances in the absorption signal. Compared to spherical shells, broken symmetry gives rise to a larger energy splitting between the lowest valence band states with dominant p-type and s-type envelope wave functions, consistent with the large Stokes shift between the PL peak and the absorption edge observed for CdS/CdSe/CdS embedded quantum shells [112, 131].

3.4
Parabolic Quantum Dot Confinement

In Section 2.3.2 we already introduced the parabolic confinement model for quantum dots with a flat shape, such as interface fluctuation quantum dots or some types of self-assembled quantum dots.

Again, the confinement along the growth direction z is very strong. Depending on the circumstances it may be convenient to assume a rectangular or a parabolic confinement potential. Along z, the wave function is considered to always be in the ground state as the level splittings are typically very large. Due to strain, the energetically lowest hole states are typically hh states.

We then introduce an additional confinement in the plane according to the shape of the quantum dot. In general, for a confinement with circular symmetry in the xy plane, the z component F_z of the angular momentum $\mathbf{F}$ is a good quantum number. Then, the confined valence band states have pure hh or lh character since they can be classified according to $\mathbf{F}^2$ and F_z. In Section 2.3.2 we used the quantum numbers l and m for the sake of a simple notation.

As an approximation for the low-lying quantum dot states, we assume a parabolic in-plane confinement with frequency ω_b,

$$V_{\mathrm{qd},b}(x,y) = \frac{m_b \omega_b}{2}\left(x^2 + y^2\right), \tag{3.56}$$

which provides the well known ground state of the envelope wave function,

$$\psi^b_{\mathrm{HO},0}(x,y) = \frac{1}{a_b \sqrt{\pi}} \exp\left[-\frac{1}{2a_b^2}(x^2 + y^2)\right]. \tag{3.57}$$

Here, we denote $a_b = \sqrt{\hbar/m_b \omega_b}$ the effective electron (hole) Bohr radius for $b = c$ ($b = v$). The energy spectrum consists of equally spaced levels separated by an energy $\hbar \omega_b$.

There are exact solutions for the eigenfunctions if a perpendicular magnetic field B is included into the two-dimensional harmonic confinement [136, 137]. Those new eigenstates are called Fock–Darwin states. Here we provide the general solution which, for example, has been applied to model the states in laterally [138, 139] or vertically [140] adjacent quantum dots. The Fock–Darwin ground state is given by

$$\psi^b_{\mathrm{FD},0}(x,y) = \sqrt{\frac{\beta_b}{\pi a_b^2}} \exp\left[-\frac{\beta_b}{2a_b^2}(x^2 + y^2)\right]. \tag{3.58}$$

Here, $\beta_b = \sqrt{1 + (\omega_b^L/\omega_b)^2}$ is a compression factor due to the magnetic field, where $\omega_b^L = eB/2cm_b$ is the Larmor frequency. The energy level spacing of Fock–Darwin states is given by $\hbar \omega_b \beta_b$. It follows from Eq. (3.58) that the orbital effect of the magnetic field, giving rise to the cyclotron motion of carriers, induces an increased confinement that is characterized by the parameter β_b, which provides a wave function compression and an increase of the level spacing. The excited Fock–Darwin states are obtained analogous to the usual harmonic oscillator states, taking into account the appropriate Hermite polynomials in x and y.

For a pair of laterally coupled quantum dots [138, 139], consisting of quantum dots $D = 1, 2$ centered at positions $\mathbf{r} = \mathbf{a}_{bD}$ that lie symmetric to the origin in the xy plane, the vector potential is taken in the symmetric gauge, $\mathbf{A}(\mathbf{r}) = (\mathbf{B} \times \mathbf{r})/2$.

Hence, the vector potential is centered at the origin, not at the individual quantum dots. This gives rise to an additional phase factor in the wave function Eq. (3.58) for quantum dot D, which then takes on the form

$$\psi_{\mathrm{FD},0}^{b,D}(x,y) = \sqrt{\frac{\beta_b}{\pi a_b^2}} \exp\left[-\frac{\beta_b}{2a_b^2}(\mathbf{r}-\mathbf{a}_{bD})^2\right] \exp\left[\frac{iq_b}{2el_B^2}(\hat{\mathbf{z}}\times\mathbf{a}_{bD})\cdot\mathbf{r}\right].$$

(3.59)

Here, $\hat{\mathbf{z}}$ is the unit vector along z, $l_B = \sqrt{\hbar c/eB}$ is the magnetic length, and q_b is the charge of the confined carrier, that is, $q_b = \pm e$ for electrons or holes, respectively, where e is the elementary charge.

While quantum dots are in reality three-dimensional objects, already a two-dimensional model, taking Fock–Darwin states as above into account provides qualitatively useful insights for the properties of coupled quantum dots [138]. However, for "flat" quantum dots that are vertically coupled, a three-dimensional model is necessary to take tunneling between the dots into account [140].

In reality, circular confinement as we assumed above, is of course, only an idealized model that neglects any anisotropies that may be present in the quantum dot shape or also the crystal lattice. An anisotropic confinement in the plane of the quantum dot, such as an elliptical shape, for example, induces mixing of angular momentum eigenstates, among them the energetically close valence band levels, such as heavy and light holes. Yet, we have focused on quantum dots with a pure hh ground state and circular in-plane confinement here, motivated by evidence of weak level mixing in many experiments.

3.5
Extensions of the Noninteracting Single-Electron Picture

Taking into account more than one confined electron in a quantum dot means that we need to carefully reconsider our model. So far, we have relied on an effective single-electron picture, where the total interaction with the crystal has – thanks to periodicity and crystal symmetry – been plugged into effective single-particle parameters, such as the effective mass. We now extend our model in order to describe interacting few-carrier states in quantum dots. Obviously it can be expected that the crystal-induced properties such as the effective mass do not change when having identical particles confined together in a quantum dot. However, we need to take into account the overall symmetry of the many-particle wave function and many-particle interactions.

3.5.1
Symmetry of Many-Particle States in Quantum Dots

If we model several carriers in a quantum dot we need first of all to take into account their quantum mechanical indistinguishability. For electrons this results in taking Fermi–Dirac statistics into account, that is, the total multi-electron wavefunction must be completely antisymmetric. We remember that for free electrons, a completely antisymmetric wave function can be formed either by an antisymmetric orbital and a symmetric spin part of the wave function, or vice versa, by a symmetric orbital part and an antisymmetric spin part. In the frame of the envelope function approximation for c band electrons, we find an analogy of the envelope function with the orbital wave function of a free electron, and of the band-edge Bloch function with the spin wave function, respectively.

For the case of two identical c band electrons, we recover the analog of the triplet and singlet states, where the spin wave functions are replaced by the corresponding Bloch functions. For the v band, the situation is a bit different due to the strong spin–orbit coupling of the bands. For degenerate hh and lh states, the Hilbert space is four-dimensional and there are wider possibilities for an antisymmetric Bloch part than for the c band. For the case of hh states with the lh states split far away in energy, however, we recover two twofold degenerate Bloch states in the absence of external magnetic fields. The symmetric and the antisymmetric Bloch states formed by pure hh states or pure lh states resemble the singlet and triplet states. However, as we know from Clebsch–Gordan theory, the antisymmetric hh Bloch state, for example, is not a total angular momentum zero or s state, but rather a superposition of an s and a d state. We therefore refer to the antisymmetric hh state as a pseudo singlet, for example, and not as a singlet.

3.5.2
Coulomb Interaction

Considering more than one electron in a quantum dot also means that we need to take the Coulomb interaction into account. The Coulomb potential describes the pairwise interaction of two particles with electric charges $q_{1,2}$ and positions $\mathbf{r}_{1,2}$, and is given by

$$V_{\text{Coul}}(\mathbf{r}_1, \mathbf{r}_2) = \frac{q_1 q_2}{4\pi\epsilon} \frac{1}{|\mathbf{r}_1 - \mathbf{r}_2|} \, , \tag{3.60}$$

with the dielectric constant $\epsilon = \epsilon_0 \epsilon_r$, where ϵ_r is the dielectric number. In principle, this takes us beyond the picture of noninteracting single particles, which we applied earlier in this chapter in the derivation of the confined states in quantum dots. The Coulomb potential enters as an additional term, possibly redefining the energy structure of confined quantum dot states.

Yet, in many types of quantum dot structures the Coulomb interaction energy between confined charge carriers is much smaller than the typical energy level splittings due to the confinement. In these cases one can take the Coulomb interaction into account by perturbation theory, for example.

In the other extreme case, that is, for quantum dots with energy level splittings that are small compared to Coulomb interaction energies, an ansatz based on noninteracting wave functions is far from the real situation because the Coulomb repulsion or attraction induces significant changes in the structure of the orbital wave functions. These considerations are also important for bound pairs of electrons and holes, the so-called excitons.

3.5.3
The Concept of Excitons in Quantum Dots

A bound pair of a c band electron and a v band hole in a semiconductor is called an exciton. Excitons can be created, for example, by optically exciting an inter-band transition in a semiconductor. The electron and the hole of an exciton usually form a bound state in the bulk crystal due to the Coulomb interaction between the negative charge $-e$ of the electron and the effective positive charge $+e$ of the hole.

Depending on the ratio of the effective masses of the electron and the hole, m_e/m_h, an exciton may be similar to an effective hydrogen atom (if the hole is much heavier than the electron, $m_e/m_h \ll 1$), or may resemble rather the positronium (for $m_e/m_h \sim 1$) and shows a corresponding characteristic spectrum of bound states in bulk semiconductors. Excitons may also form complexes of several excitons that are bound together, so-called multi-excitons.

The three-dimensional confinement of quantum dots usually alters some properties of excitons when compared to free excitons in a bulk semiconductor, due to the additional boundary conditions imposed on the electron and hole wavefunctions. Reducing the spatial extension of quantum confinement leads in general to energy level shifts toward higher energies, as the energy of the confined states increases. This spectral blue shift is sometimes called the confinement effect. Further, if the confinement geometry imposes a reduced symmetry on the confined states, then qualitative changes may be observed in the optical spectra, such as the level mixing and redistribution of spectral weights mentioned in Section 3.3.3, or the fine structure splitting of the exciton levels in quantum dots with a slightly elliptical shape mentioned in Section 6.2. Such effects are basically due to state mixing and may be observed in the spectra, as radiative transitions among the confined quantum dot levels satisfy certain selection rules, which are discussed in more detail in Chapter 5.

When observing the spectra of exciton and multi-exciton states in quantum dots, we observe not only features due to the quantum confinement, but also due to the Coulomb interaction. The Coulomb interaction already plays a role on the single-exciton level, as there is a competition of the Coulomb interaction versus the quantum dot confinement in defining the states and the energy spectrum. In this competition, the relative magnitude of the associated energies is essential. In the so-called strong-confinement regime, the single-particle confinement energies due to the quantum dot potential are much larger than the electron–hole Coulomb interaction energy. The exciton energy spectrum is then determined mainly by the confinement. An alternative definition of the strong-confinement regime can be given as the limit in which the quantum dot radius is much smaller than the "natural" Bohr radius a_X of a free exciton. In this book, we focus on quantum dots in the strong confinement regime. In the strong confinement regime we usually make an ansatz for excitons with noninteracting single-particle wavefunctions for electrons and holes and take the Coulomb interaction into account, for example, in the frame of perturbation theory, or in a variational ansatz.

3.5.4
Carrier Configurations in the *s* Shell and Energies

Here we briefly return to the parabolic quantum dot model and look at the different *s*-shell configurations. These configurations are of particular interest for many proposals to process quantum information with exciton or spin qubits in quantum dots. In the low-excitation regime, when optically active quantum dots are in equilibrium occupied by no or one electron and/or hole, carriers relax to the *s*-shell typically on a sub-ns timescale from the higher $p, d, \ldots$ levels.

The four optically active *s*-shell states are depicted schematically in Figure 3.2, which shows the level diagrams for electrons (▲) and holes (△), analogous to Figure 2.9 with the spin orientation indicated by the orientation of the symbols. Figure 3.2a–d shows the neutral exciton (X^0), the negatively and positively charged excitons (X^{1-}, X^{1+}), where one additional carrier joins the exciton, and the biexciton ($2X^0$). In addition to the states shown for X^0, X^{1-}, and X^{1+}, there is a second version with inverted spins, giving rise to an inverted polarization of the emitted photon. The optical selection rules for the recombination process are discussed later in Section 5.2.2. In addition to these optically active or "bright" excitonic states, there are also two optically inactive or "dark" exciton states in the *s* shell. These are obtained from the state shown in Figure 3.2 for X^0 when inverting either the electron or the hole spin. This configuration leads to an angular momentum difference of ± 2 for the transition, which cannot be achieved with an electric dipole transition, hence these states are "dark". In the following we focus on the "bright" excitonic states.

The charged excitons X^{1-} and X^{1+} are special as they have a filled *s* shell for one carrier type, respectively. The *s* orbital state of the *c* band can be occupied by maximally two electrons, forming a singlet state. Electrons with equal spins cannot occupy the *s* shell due to the Pauli exclusion principle. For the *s* shell of the *v* band we recall from our discussions earlier in this chapter that in self-assembled quantum dots the hh and lh states are typically split such that the *s* shell provides

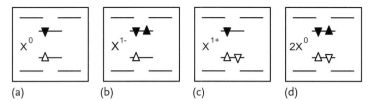

(a)   (b)   (c)   (d)

Fig. 3.2 The four types of bright *s*-shell excitonic states in a quantum dot: (a) the neutral exciton X^0, the (b) negatively and (c) positively charged exciton, X^{1-} and X^{1+}, respectively, and (d) the biexciton $2X^0$. For the first three, there is also a spin-inverted version not shown in this figure.

two hh states. Again, if two holes are present, they have an antisymmetric angular momentum (Bloch) part that we refer to as a pseudo singlet.

Taking into account the Coulomb interaction for electrons and holes, we can calculate the energies of the above states. We assume the strong confinement regime and take the Coulomb interaction as a perturbation of the confined states into account to first order. The obtained energies are

$$E_{X^0} = E^e + E^h - V^{eh} \,, \tag{3.61}$$

$$E_{X^{1+}} = E^e + 2E^h + V^{hh} - 2V^{eh} \,, \tag{3.62}$$

$$E_{X^{1-}} = 2E^e + E^h + V^{ee} - 2V^{eh} \,, \tag{3.63}$$

$$E_{2X^0} = 2E^e + 2E^h + V^{ee} + V^{hh} - 4V^{eh} \,, \tag{3.64}$$

where E^e and E^h are the confinement energies of electrons and holes, respectively, and the Coulomb interaction energies are given by

$$V^{eh} = \frac{e^2}{4\pi\epsilon} \langle \psi_e \psi_h | \frac{1}{|\mathbf{r}_e - \mathbf{r}_h|} | \psi_e \psi_h \rangle \,, \tag{3.65}$$

$$V^{ee} = \frac{e^2}{4\pi\epsilon} \langle \psi_{e1} \psi_{e2} | \frac{1}{|\mathbf{r}_{e1} - \mathbf{r}_{e2}|} | \psi_{e1} \psi_{e2} \rangle \,, \tag{3.66}$$

$$V^{hh} = \frac{e^2}{4\pi\epsilon} \langle \psi_{h1} \psi_{h2} | \frac{1}{|\mathbf{r}_{h1} - \mathbf{r}_{h2}|} | \psi_{h1} \psi_{h2} \rangle \,. \tag{3.67}$$

3.6
Few-Carrier Spectra of Self-Assembled Quantum Dots

We return here to the self-assembled quantum dots introduced in Section 2.3 and apply the theoretical considerations of the present chapter to this system.

3.6.1
From Ensemble to Single Quantum Dot Spectra

In the ensemble spectra of self-assembled quantum dots, such as Figure 2.8, a clear shell structure is revealed. The spacing between the different shells is on the order of several tens of meV. We have mentioned in Section 2.3.2 that the shell structure of these dots can be approximated well by a harmonic oscillator model for both electrons and holes. From this model we can now conclude how many carriers can be accommodated in each shell. Further, they can be filled sequentially when increasing the optical excitation density, starting from the lowest *s*-levels for electrons and holes.

Within a quantum dot ensemble, each individual dot has its distinct energy structure. However, the expected sharp emission lines from single-dot transitions remain hidden due to the random size and morphology distributions in the ensemble. A solution to circumvent this averaging problem is the isolation of individual dots. This can be achieved, for example, by using the masking technique described in Chapter 2, or the fabrication of low surface-density ensembles by stopping the substrate rotation during quantum dot growth. To be able to repeatedly study the same dot over a longer period of time, a masking technique can be applied to find the dot of interest again. Using such techniques it is possible to isolate single self-assembled dots and to study the structure of the different orbital shells in more detail.

In Figure 3.3, PL signals recorded of a single dot are shown. These spectra show narrow lines with width $\Gamma < 150\,\mu eV$, which can be attributed to different occupancy states of the dot. These states as well as their coherent properties have been investigated in great detail by many researchers [22, 27, 29, 42, 141–143]. Single quantum dots have been studied especially by power dependent PL spectroscopy and often in combination with applied static electric fields.

The most simple occupancy state giving rise to an optical transition is the single-exciton state X^0. Figure 3.3 shows emission spectra of a single dot as a function of excitation power. At low power only one line is detected, originating from the decay of X^0. With increasing excitation power, a second line appears in the s-shell with a quadratic dependence on excitation density. This line is attributed to the biexciton $2X^0$, where the quantum dot is occupied by two electron–hole pairs in the s shell. To be precise, the observed additional line results from the optical decay of

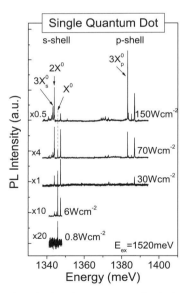

Fig. 3.3 Photoluminescence of a single quantum dot as a function of excitation power. At low powers only the two s-shell transitions are observed and with increasing powers also p-shell transitions are observed. Data courtesy J. J. Finley, Walter Schottky Institute, TU München.

one electron–hole pair of the biexciton, leaving behind the second one (which itself would then decay as a X^0). The $2X^0$ transition energy is shifted to lower energy with respect to the X^0 transition due to the additional many-particle interactions of the biexciton. We elaborate on these interaction-induced shifts in Section 3.6.2. At even higher excitation powers, also the p-shell is occupied, which, due to additional interactions with additional carriers, leads to further shifts, such as the new s-shell transition $3X^0_s$. These observations match with the previously described model of a quantum dot shell structure for electrons and holes when few-carrier interactions are taken into account. a The observed inter-shell spacing for this type of dot is typically on the order of 20–50 meV and provides a relatively large energy scale when compared to the interaction-induced line shifts which are in the 1–10 meV range.

3.6.2
Transition Energies of Few-Particle States

We have given the energies of the excitonic states of the s-shell above in Section 3.5.4. These are straightforward to expand by including the p and higher shells. Instead of proceeding with this here, let us now look in more detail at the transition energies from the s-shell states. Due to energy conservation, the observed photon energy ΔE is given by the difference between the initial and final excitonic states,

$$\Delta E = E^{\text{in}} - E^{\text{fin}}. \tag{3.68}$$

If the initial state is X^0, X^{1+}, or X^{1-}, the final state is just the crystal ground state (with energy $E_{\text{cgs}} = 0$), the single-hole state (with energy E_h), or the single-electron state (with energy E_e), respectively. For these transitions, the transition energies according to Eqs. (3.61)–(3.63) are given by

$$\Delta E_{X^0} = E^e + E^h - V^{eh}, \tag{3.69}$$

$$\Delta E_{X^{1+}} = E^e + E^h + V^{ee} - 2V^{eh}, \tag{3.70}$$

$$\Delta E_{X^{1-}} = E^e + E^h + V^{hh} - 2V^{eh}. \tag{3.71}$$

Here, we notice that if there is a difference in the absolute magnitude of the Coulomb interaction between like (V^{ee}, V^{hh}) and unlike (V^{eh}) particles, the transition lines of charged excitons will shift in the PL spectrum with respect to the neutral exciton line. Since the hole wave function is typically more strongly localized than the electron wave function, the ordering of the Coulomb matrix elements can sometimes be assumed as $V^{ee} < V^{eh} < V^{hh}$. Based on these relations, we can estimate the shifts of the s-shell charged excitons with respect to X^0: Since $V^{ee} < V^{eh}$, we expect that X^- is shifted to lower energies, while due to $V^{eh} < V^{hh}$, X^+ is shifted to higher energies. The resulting line ordering is most of the time observed for self-assembled QDs but does not necessarily hold true for any QD system since it depends on the particular Coulomb interactions.

Obviously, if the final state contains more than a single particle, the complete interaction energy of the final state must be taken into account as well. For example, if we look at the decay of a biexciton into an exciton, we obtain from Eqs. (3.61) and (3.64),

$$\Delta E_{2X^0} = E^e + E^h + V^{ee} + V^{hh} - 3V^{eh} . \tag{3.72}$$

This transition energy is different from the exciton transition Eq. (3.69) if $V^{ee} + V^{hh} \neq 2V^{eh}$. Similarly as for the charged excitons, the biexciton transition line may be shifted to lower or higher energies in the PL spectrum, depending on the ordering of the interaction energies between electrons and holes. In Figure 3.3, the biexciton line is shifted to lower energies, indicating that the attractive interaction between electrons and holes overcompensates the repulsion of identical carriers

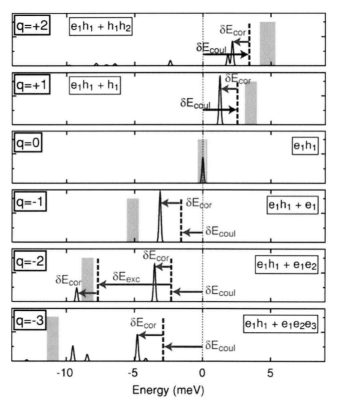

Fig. 3.4 Calculated transition energies for neutral and charged excitons in a single quantum dot relative to X^0 (e_1h_1). The arrows show the different contributions to the total energy. The Coulomb interaction (δE_{coul}) leads to shifts to lower and higher energy, and correlation effects (δE_{cor}) only to lower energy. Reprinted with permission from [144]. Copyright (2001) by the American Physical Society.

in the dot under study. The shift of the biexciton relative to the exciton transition energy, $\Delta E_{X^0} - \Delta E_{2X^0}$, is called the biexciton binding energy. For most quantum dot systems, a positive biexciton binding energy was observed, that is, a shift of the biexciton to lower energies compared to X^0 [22, 143]. This behavior is also obtained

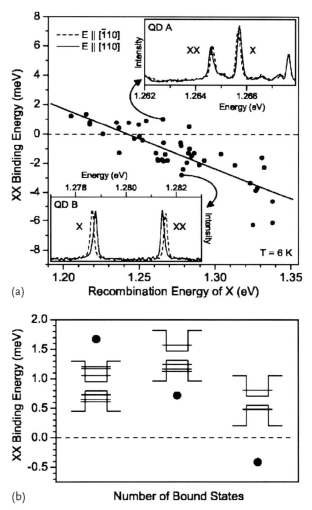

Fig. 3.5 (a) Observed biexciton binding energy (symbols) as a function of quantum dot transition energy. A transition from "binding" to "anti-binding" biexcitons is observed for increasing transition energy. (b) Calculated binding energy as a function of bound states reproduces the observed behavior. Reprinted with permission from [145]. Copyright (2003) by the American Physical Society.

theoretically for a harmonic oscillator potential, taking experimental values for the effective masses into account [83].

Thus, the shifts observed with transition energies when additional carriers are introduced provide a measure for the difference of few-particle Coulomb energies.

In the discussion of the excitonic emission of a quantum dot we have accounted so far only for the single particle energies of the different carriers and their mutual direct Coulomb interaction in first-order perturbation theory. For a more precise description, the Coulomb exchange interaction and also higher-order correlations need to be taken into account.

Figure 3.4 shows the results of a combined theoretical and experimental study of charged excitons in a single quantum dot by Regelman and coworkers [144]. The calculated emission intensity (solid lines) is plotted as a function of energy relative to the neutral exciton transition. The labels give the number of additional carriers compared to the single exciton ($e_1 h_1$). A comparison of the calculated emission energies with the experimental result (shaded bars) shows good qualitative agreement. The calculations by Regelman *et al.* confirm that the Coulomb interaction can induce shifts (labeled δE_{coul}) to lower and higher energies, whilst correlation shifts (δE_{cor}) occur only toward lower energies. In addition, as discussed in the previous paragraph, the Coulomb interaction depends strongly on the spatial extension of the carrier wavefunctions. Therefore, strong variations of the magnitude of these shifts are observed when comparing different quantum dots due to morphology fluctuations. In contrast, correlation effects lead to approximately constant shifts to lower energy.

Figure 3.5 shows results of a combined experimental and theoretical study by Roth *et al.* [145] of the biexciton binding energy in single quantum dots as a function of the observed X^0 energy. In Figure 3.5a the experimentally observed biexciton binding energy, $\Delta E_{X^0} - \Delta E_{2X^0}$, is plotted as a function of the exciton energy. A clear transition from "binding" ($\Delta E_{X^0} - \Delta E_{2X^0} > 0$) to "anti-binding" ($\Delta E_{X^0} - \Delta E_{2X^0} < 0$) biexcitons are observed for decreasing dot size, that is increasing transition energy. This observation was explained by a strong variation of the correlation contribution δE_{cor} for the biexciton state. A comparison of experimental findings to theoretical calculations of the correlation effects showed that for larger dots, mostly "binding" biexcitons are expected, whereas for smaller dots correlation effects are less pronounced and "anti-binding" biexcitons can be observed. This behavior is attributed to a decrease in the number of bound states in the dot as demonstrated in Figure 3.5b [145, 146]. As a side remark, one should exercise care with the terminology of "binding" and "anti-binding" biexcitons since all particles are bound by the quantum dot confinement, which is in contrast to the bulk crystal, where "anti-binding" biexcitons would split into two excitons.

4

Integration of Quantum Dots in Electro-optical Devices

The ability to manipulate and control the properties of quantum dots and other nanostructures is a key requirement for any application. We have shown in the previous chapters that the electronic and optical properties are determined by the structural, morphological and chemical properties of the dot, which can be controlled only to a certain degree. Therefore, a knob to turn is required to tune the electronic levels, add or remove charges to or from the dot or even turn on and off interactions between two neighboring dots. An elegant parameter for this purpose is a static electric field. We will show that in a semiconductor diode such an electric field, and also the number of particles in a dot, can be adjusted and controlled by simply changing an applied gate voltage. Furthermore, the optical response of a QD can be strongly modified by the optical field of a cavity, which we will introduce in the last part of the chapter.

4.1
Tuning Quantum Dots by Electric Fields

The number of carriers injected into a QD from a doped contact can be adjusted deterministically via Coulomb blockade [40, 147], furthermore, the energy levels for both electrons and holes can be tuned. This results in a shift of the optical transition frequency due to the quantum confined Stark effect (QCSE). The QCSE can be further used to turn on and off quantum mechanical coupling in a pair of QDs providing the basis for a scalable qubit architecture. In this section we introduce the most frequently used device, which allows for both the application of static electric fields and control of the number of carriers confined in a QD or pair of coupled QDs.

4.1.1
Semiconductor Diodes

The most frequently used devices to apply static electric fields and control the number of carriers (electrons or holes) in epitaxially grown QD nanostructures by a gate voltage are semiconductor diodes. These can be either bipolar *p-i-n* structures or

Spins in Optically Active Quantum Dots. Concepts and Methods.
Oliver Gywat, Hubert J. Krenner, and Jesse Berezovsky
Copyright © 2010 WILEY-VCH Verlag GmbH & Co. KGaA, Weinheim
ISBN: 978-3-527-40806-1

unipolar metal semiconductor junctions. The latter type of device are so-called Schottky diodes and are most frequently used if only one carrier species is to be electrically controlled. If electrons or holes are to be injected in the QDs either *n-i-* or *p-i-*Schottky diodes are used. If both carrier species are required, for example, for QD-lasers or single photon sources *p-i-n* diodes are the devices of choice [148–150].

The band structure of an *n*-type Schottky diode is schematically depicted in Figure 4.1. In this case by changing the gate voltage (V_{Gate}) *electrons* can be added from the *n*-region to the QD and simultaneously a *positive* static electric field is tuned over the QD layer. For *p*-type devices the electric field orientation is reversed and holes can be loaded into the QDs [151]. In the following we restrict the discussion to *n*-type devices although the same arguments apply also for *p*-contacts and holes taking into account the inverted polarity.

As we can see from Figure 4.1 the QDs themselves are embedded in the intrinsic (unintentionally doped) region of length $d_{intrinsic}$ of the junction and are separated by a tunneling barrier of thickness d_{tunnel} from the heavily doped *n*-contact. The *i*-region above the QDs typically contains a higher bandgap AlGaAs barrier or short-period AlAs/GaAs superlattice to prevent current flow and leakage in forward direction. Since the Fermi energy, E_F, is pinned close to the conduction band edge in the *n*-contact, the electric field F drop across the intrinsic region can be tuned by applying a gate voltage V_{Gate} between the *n*-contact and a top metal electrode forming a Schottky-contact. For heavily doped contacts and long intrinsic regions,

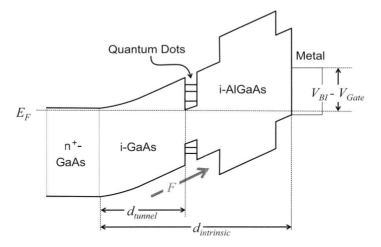

Fig. 4.1 Band structure of an *n-i* Schottky diode with QDs embedded in the intrinsic region. In such a device the static electric field F can be tuned by applying a gate voltage (V_{Gate}) between the doped *n*-contact and the metal Schottky gate. Furthermore, a well defined number of carriers can be injected across a tunneling barrier from the contact into the QDs.

F can be approximated linearly as

$$F = \frac{V_{BI} - V_{Gate}}{d_{intrinsic}} , \tag{4.1}$$

with V_{BI} being the built-in voltage, that is, the "height" of the Schottky barrier. For GaAs-based structures V_{BI} is typically in the range of $\sim$0.8 V and only weakly dependent on the metal used for the gate electrode. In contrast, for *p-i-n* junctions V_{BI} is given by the energy differences between the hole and electron quasi Fermi levels in the *p* and the *n* regions. For high doping concentrations these are close to the band edges, giving rise to a built-in potential of approximately the bandgap of the material. In typical GaAs-based devices electric fields $|F| > 300\,$kV/cm can be applied before electrical breakdown occurs in the reverse direction. Such a static electric field provides a powerful tuning mechanism since it can be used to tune the energies of optical transitions in QDs via the dc Stark effect (QCSE) which, furthermore, can be used to manipulate the interdot couplings in pairs of closely stacked QDs.

4.1.2
Voltage-Controlled Number of Charges

In an epitaxially grown structure, the layer thicknesses can be adjusted with mono-layer precision, allowing for delicate tuning of the device properties. As already indicated in Figure 4.1, the distance between the *n*-contact and the QD layer(s), d_{tunnel} can be precisely tuned, which controls the interaction and tunneling rate between electrons in the reservoir and confined in the QD.

In particular, for small values of d_{tunnel} only a short triangular barrier exists through which carriers can efficiently tunnel from the contact acting as a reservoir into the QDs. This process is schematically depicted in Figure 4.2 for the case of a single QD. Under reverse bias ($V < V_{1e}$) (see Figure 4.2a), the Fermi energy of the electron reservoir is below the lowest energetic electron level (*s*-shell) of the QD. Under these conditions no electron tunnels from the contact into the QD, which remains empty. As the forward bias is increased to $V_{1e} \leq V \leq V_{2e}$ the *s*-states of the QD shift below E_F, and one electron can enter the QD as shown in Figure 4.2b. Tunneling of a second electron is blocked due to the Coulomb repulsion by the electron occupying the dot. This effect is referred to as the Coulomb blockade and is often described by considering the QD as a small capacitor with a capacitance (C_{QD}). In this picture, which is often applied for electrostatically de-fined QDs [20, 86], the energy level of the second electron or hole to be added into the QD is shifted toward higher energies by the Coulomb addition energy depending only on the elementary charge and the capacitance of the QD,

$$E_C = \frac{e^2}{C_{QD}} . \tag{4.2}$$

As a direct consequence, this additional Coulomb energy contribution has to be overcome, and therefore, this effect is commonly referred to as the Coulomb block-

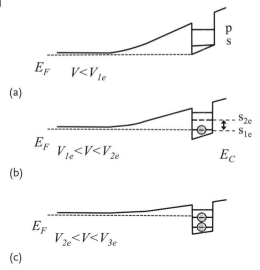

E_F $V<V_{1e}$

(a)

E_F $V_{1e}<V<V_{2e}$ E_C

(b)

E_F $V_{2e}<V<V_{3e}$

(c)

Fig. 4.2 Deterministic charging of a QD via Coulomb block-ade. (a) For low voltages the QD is empty. (b) When the s-level of the QD drops below the Fermi energy, one electron enters the QD. (c) The second electron can only enter the QD when the forward voltage is increased further to compensate for the Coulomb addition energy.

ade. We want to note that this effect is observed in any QD-like system such as, for example, gate-defined QDs at sufficiently low temperatures with $k_B T < E_C$. On-ly if the forward bias is increased even further to $V > V_{2e}$ is a second electron transferred into the QD.

Adding charge carriers into the QD gives rise to a pronounced response of the device capacitance [147]. At each voltage where the QD occupancy changes by one carrier, a characteristic step or peak occurs in the capacitance-voltage trace allowing for precise measurements of the Coulomb addition energies. A typical example is shown in Figure 4.3a for a hole (*p*-type) and 4.3b electron (*n*-type) charging sample. Increasing positive (negative) gate voltage corresponds to forward biasing the *n*- (*p*)-type device, which allows for the addition of electrons (holes) to the QDs. For both the electron and hole charging samples, distinct charging features are resolved as the junction is brought toward forward bias. These peaks correspond to filling the ground (*s*-shell) with one and two electrons/holes and then the first excited state (*p*-shell) of the QDs with increasing forward bias. For all shells, distinct peaks can be identified, each of which correspond to an increase of the number of carriers in the QD by 1. From these charging experiments detailed information on the electronic level structure can be deduced. For example, it has been shown that for holes, Hund's rule and the Aufbau principle of atomic physics are violated [152–154].

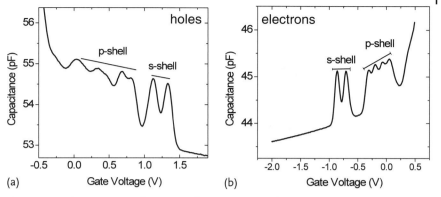

Fig. 4.3 Typical capacitance-voltage traces for a *p*-type (a) and *n*-type (b) Schottky diode that allows for adding electrons and holes into the QDs' *s*- and *p*-shells, respectively. Data courtesy Andreas D. Wieck, Ruhr-Universität Bochum.

4.1.3
Optically Probing Coulomb Blockade

In optically active QD structures the Coulomb blockade can be observed using optical spectroscopy, in addition to electrical means. By adding a photogenerated electron–hole pair, a single exciton ($X^0 = 1e + 1h$), the emission of this exciton, and any electrically injected excess charges can be probed [40, 155].

In Figure 4.4 we present a typical example of a photoluminescence experiment performed on a single self-assembled InGaAs QD embedded in a charge-tunable *n-i* Schottky diode. The optical transitions attributed to the main emission lines are shown in Figure 4.5. As the gate voltage is tuned, a clear switching behavior between different emission lines is observed, which can be explained by discrete charging events. Moreover, all spectral lines show a pronounced energy shift as the gate voltage is tuned. This shift is due to the QCSE since the electric field is also tuned.

For large reverse bias, $V < V_1$, no excess electrons are injected from the *n*-contact. Therefore, a single emission line of the optically added exciton X^0 is detected. As the reverse bias is decreased to $V_1 < V < V_2$ a first electron tunnels into the QD. The presence of this electron leads to the additional formation of singly negatively charged excitons, $X^{1-} = 2e + 1h$ in this voltage range. The emission line of X^{1-} is significantly red-shifted by ~4.7 meV with respect to X^0. This shift occurs due to the modified few-particle Coulomb interactions (*e–e* repulsion and *e–h* attraction), which we discuss in Section 3.5. Moreover, this shift typically ranges in the order of 3–6 meV for most commonly studied QD systems.

For $V_2 < V < V_3$ the *s*-shell of the QD gets populated by a second electron by injection from the doped reservoir. However, in contrast to the neutral and singly charged excitons, the addition of a second electron gives rise to *two* new emission lines labeled X^{2-}_{singlet} and X^{2-}_{triplet}. From the gate voltages at which switching

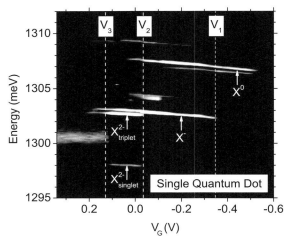

Fig. 4.4 Charged exciton emission from a single QD. For increasing V_G electrons are controllably added to the QD and emission of the neutral (X^0), singly (X^{1-}) and doubly negatively (X^{2-}) charged excitons is detected.

between different exciton species occurs the Coulomb addition energy can be deduced analogous to purely electrical techniques. Similar data has been obtained by several groups worldwide demonstrating the general nature of the underlying mechanisms [40, 151, 156–162].

The observation of two spectral lines for X^{2-} can be explained by taking into account both the initial and final state of the radiative decay and, moreover, the exchange interaction between the spins. The radiative decays of the four emission lines detected are summarized in Figure 4.5. For X^0 and X^{1-} only the *s*-shell is occupied in the initial and final state of the decay. The shift between the two emission lines arises from the additional *direct* Coulomb interaction terms between the electron(s) and the hole in the initial state. For X^{1-} the *s*-shell in the conduction band is filled completely with two electrons of opposite spin orientation and, therefore, additional contribution from *exchange* Coulomb interactions do not have to be taken into account.

In contrast, for X^{2-} the optically generated electron cannot relax to the filled *s*-shell of the QD but remains in a *p*-level with random spin orientation (see the lower panel of Figure 4.5). When the hole recombines with one of the two electrons confined in the *s*-shell one electron remains in both the *s*-shell and the *p*-shell. Since these two electrons occupy different orbital states, their spin wavefunctions can form both singlet $S = 0$ or triplet $S = 1$ states. Due to the electron *s–p* exchange interaction the absolute energy of the triplet state is shifted to lower energy relative to the singlet level. In the optical decay the energy difference between the initial X^{2-} and the final $2e$ state is measured, and therefore, the emission line of the higher lying singlet state is shifted to lower energies compared to the triplet line, as observed in experiment. Moreover, for the $S = 1$ triplet three states with

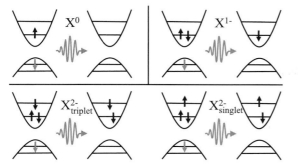

Fig. 4.5 Decay of neutral (X^0), singly charged (X^{1-}) and doubly charged exciton (X^{2-}). For X^{2-} two final states exist for which the two electrons in the s and p shell form either a spin singlet or triplet which gives rise to the observation of two emission lines.

$M_S = 0, \pm1$ whilst for the $S = 0$ singlet only one state with $M_S = 0$ exists. This ratio of 3 : 1 is also reflected in the experimental data shown in Figure 4.4 by the intensity ratio of the two spectral lines. These findings clearly indicate that direct and exchange few-particle Coulomb interactions give rise to pronounced shifts and their contributions can be clearly identified in the optical spectrum. Moreover, for optical transitions both the initial *and* final state have to be taken into account. In particular for the coupled QD structures discussed in Chapter 8 both aspects are crucial to explain the complex emission patterns.

Finally, we want to note that the data shown in Figure 4.4 was recorded for a rather long barrier between the *n*-contact and the QDs of 40 nm. For such long barrier widths the injection rate is low and the QD is not in a charge equilibrium giving rise to the observed parallel emission of differently charged excitons at the same gate voltage. When d_{tunnel} is reduced, the tunneling time for electrons in and out of the QD can be tailored, and discrete charging steps can be resolved [158]. However, short tunneling times can lead to a significant reduction of the coherence time of spins and excitons due to interaction via tunneling or scattering between the confined spin in the QD and the (mostly) incoherent bath of spins in the *n*-contact. These cotunneling effects have to be minimized for an efficient spin pumping scheme [163]; this is described in more detail in Section 7.1.2.

4.1.4
Quantum Confined Stark Effect

The response of quantum states to static electric fields is known as the Stark effect. For excitons confined in semiconductor nanostructures the Stark shift can exceed the exciton binding energy. This was demonstrated for the first time for quantum well excitons by D. A. B. Miller *et al.* in 1984 [48, 164]. They observed shifts much larger than the exciton binding energy, up to electric fields 50 times larger than the classical ionization field. Since this effect is a direct consequence of the quantum confinement it is referred to as the quantum confined Stark effect (QCSE). The QCSE turned out to be an extremely useful mechanism to tune optical transitions in low-dimensional semiconductor structures. Therefore, it found application in the spectroscopy of single and coupled QDs [70, 156, 165–169], since it provides extremely high accuracy simply by modulating a gate voltage.

In this section we develop a simple model of the QCSE and show some implications on the optical properties of quantum wells and QDs.

General Aspects

An exciton is formed by two particles of opposite charge, one electron and one hole, which themselves are influenced by an electric field, due to their charge. An electric field leads to a small relative displacement of e and h and a finite dipole moment. This dipole moment couples to the electric field and the Stark shift of a small excitonic dipole is given by

$$\Delta E_{\text{Stark}} = -\mathbf{p} \cdot \mathbf{F} . \tag{4.3}$$

Here $\mathbf{p}$ is the excitonic dipole moment (oriented from the positive to the negative charge) and $\mathbf{F}$ the static electric field. According to this relation the Stark effect always leads to a shift of the exciton transition to lower energy. However, at zero electric field the center of the electron and hole wavefunctions in a QD can be displaced by a distance $\mathbf{s}_0$, leading to a zero-field dipole moment $\mathbf{p}_0 = e\mathbf{s}_0$. In a first approximation, an electric field induces a small dipole moment following a linear dependence

$$\mathbf{p} = \beta \mathbf{F} + \mathbf{p}_0 . \tag{4.4}$$

Here β is the polarizability of the exciton, a measure for how easily the electron and hole can be pulled apart, that is, polarized. The resulting electron–hole separation $\mathbf{s}_{\text{ind}} = \beta \mathbf{F}/e$ is, therefore, defined through

$$\mathbf{p} = e\mathbf{s}_{\text{ind}} + \mathbf{p}_0 . \tag{4.5}$$

The magnitude of the QCSE can be derived to be

$$E = E_0 - \mathbf{p}_0 \cdot \mathbf{F} - \beta F^2 , \tag{4.6}$$

where E_0 is the zero-field exciton energy and $F = |\mathbf{F}|$. The electric field leads to a quadratic shift of the exciton transition to lower energy. The vertex of the parabola can be shifted from $\mathbf{F} = 0$ if a finite e–h displacement is present without an electric field applied. For a symmetric structure $\mathbf{p}_0 = 0$, however, as discussed later, for self-assembled QDs nonzero $\mathbf{p}_0$ is observed. In semiconductor nanostructures, the e–h displacement is limited by the size of the system. Therefore $\mathbf{s}_{\text{ind}}$ increases with $\mathbf{F}$ up to the dimension of the system and thereafter remains constant for higher fields. As a direct consequence the QCSE changes from a quadratic to a linear dependence for high electric fields. This effect can be observed only for excitons confined in low-dimensional systems since the confinement potential inhibits field ionization and

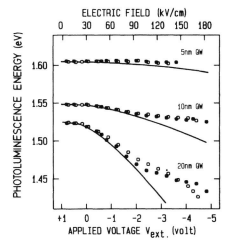

Fig. 4.6 Quantum confined Stark effect of AlGaAs-GaAs QWs of different width. The polarizability β – that is, the curvature of the Stark parabola – increases for thicker wells. Furthermore, at high fields a transition to a linear regime is observed as the wavefunctions of electron and hole are pulled towards the opposite sides of the QW. Reprinted with permission from [170]. Copyright (1985) by the American Physical Society.

increases the exciton binding energy[3]. Typical examples of the QCSE are shown in Figure 4.6 for three quantum wells of different width. For low electric fields, $F < 90 \, \text{kV/cm}$ the shift of the PL peak energy (symbols) is well reproduced by the anticipated quadratic dependence (line). The dependence weakens to linear for high F with a larger slope for the wider wells. This is also consistent with the distance of centers of the two wavefunctions asymptotically approaching the well width.

The polarizability β strongly depends on the size of the nanostructure. For a quantum well structure with infinite confinement potential the polarizability can be calculated by perturbation theory [50, 171, 172] if the electrostatic contribution is small compared to the sub-band energies of electron and hole,

$$|eFw| \ll \frac{\hbar^2 \pi^2}{2m^* w^2} \, , \tag{4.7}$$

where w is the width of the well. The Hamiltonian of the problem can be written as

$$H = H_0 + eFz \, , \tag{4.8}$$

3) For example, the binding energy increases by a factor of 4 for QW excitons compared to bulk semiconductors. These effects become more pronounced when the size of the nanostructure is smaller than or of the same order as the 3D exciton Bohr radius a_B^*. For III–V semiconductors this value is in the 10 nm range, for instance, for bulk GaAs $a_B^* \sim 11 \, \text{nm}$ [56].

where H_0 is the usual quantum well Hamiltonian with the corresponding eigenstates ψ_n. For the ground state ψ_1, which has even parity, the first order correction given by

$$\Delta E^1 = \langle \psi_1 | eFz | \psi_1 \rangle \tag{4.9}$$

equals zero. Therefore, the second-order perturbation has to be included and it can be shown that the resulting Stark shift for the interband transition between the lowest electron and hole levels is given by

$$\Delta E_{\text{Stark}} = \beta F^2 = \frac{1}{24\pi^2} \left(\frac{15}{\pi^2} - 1 \right) \frac{(m_e^* + m_h^*)e^2 w^4}{\hbar^2} F^2 . \tag{4.10}$$

In this expression $m_{e/h}^*$ are the effective masses of electron and hole. Furthermore, in this approximation we also neglected any effects of the electric field on the exciton binding energy, which is definitely required for a more realistic description. Clearly, due to the dependence to the fourth power, the polarizability β of the exciton strongly depends on the size w of the nanostructure. For example, a reduction of the effective size by a factor of 2 leads to a reduction of β by a factor of 16. Therefore, the polarizability is a very sensitive probe of the confinement potential and size of nanostructures and thus can be used, for instance, to determine the relative size of two QWs. This can be clearly seen in Figure 4.6 where the magnitude of the QCSE increases along with the well width. Furthermore, due to the strong morphological anisotropy of self-assembled QDs, a strong dependence of the QCSE on the electric field direction is expected. In the following subsection we present the structural properties of self-assembled QDs and their effect on the magnitude of the QCSE depending on the orientation of the electric field vector.

The Quantum Confined Stark Effect in Quantum Dots

In the general discussion of the QCSE in the previous section we demonstrated that the magnitude of the QCSE is strongly influenced by the size of the QD. In Figure 4.7 the energy of the s-shell transition of InAs self-assembled QDs is plotted as a function of an external electric field. In the inset the orientation of F with respect to the QD is indicated as arrows for the two different polarities of the devices used in the experiments. In contrast to typical III–V QWs the vertex of the Stark parabola is shifted to negative electric fields for self-assembled QDs, indicating a nonzero value of p_0 and, therefore, a finite displacement of the centers of gravity of the e and h wavefunctions at $F = 0$ [165]. From the electric field at which the maximum transition energy is observed, a permanent dipole moment of $p_0 \sim 7 \times 10^{-29}$ Cm can be derived. This dipole moment corresponds to an e–h displacement of $s_0 \sim 4$ Å with the hole wavefunction localized at the QD apex and the electron wavefunction at the base. This finding can be explained by an increasing In-content towards the apex of the dot. Electron states are delocalized over the whole QD due to their smaller effective mass and are less sensitive to the morphology of the island since the conduction band wavefunctions in III–V materials are affected primarily by hydrostatic components of the strain. In contrast,

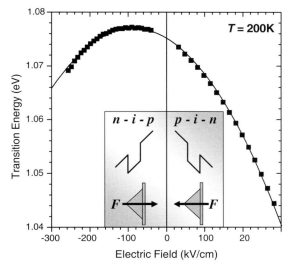

Fig. 4.7 Quantum confined Stark effect of self-assembled InAs QDs. The maximum of the parabola is shifted to negative electric fields due a finite excitonic dipole moment present for $F = 0$. Reprinted with permission from [173]. Copyright (2005) by IOP Publishing Ltd.

the higher effective mass of holes leads to a localization in areas of deeper effective confinement in regions of high biaxial strain at the apex of the QD. As a direct consequence, in the exciton the centers of the electron and hole wavefunctions are shifted apart with the hole localized on top of the electron. This results in a small permanent dipole moment of the exciton with no electric field applied.

The QCSE of a QD exciton can be directly used for laser absorption spectroscopy with high spectral resolution. Here the QD exciton energy is tuned in and out of resonance with a narrow band laser via the QCSE allowing a highly sensitive lock-in detection of the direct absorption of the QD. Due to the small energy shifts required in this technique the QCSE can be approximated linearly and for typical field-tunable structures as described in this chapter. Typical shifts are in the order of ~2.5–3 μeV per mV gate voltage applied to the device [167]. This technique will be discussed further in Chapter 7.

4.2
Optical Cavities

The interaction between light and quantum dots can be enhanced by placing the dots within an optical cavity. This section will outline several types of cavities that are readily integrated with quantum dot systems. We will also discuss a few ex-

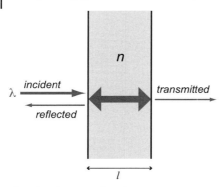

Fig. 4.8 Schematic of a Fabry–Pérot resonator consisting of two reflective surfaces surrounding a medium with index of refraction, *n*, separated by a distance, *l*. Incident light of wavelength λ is partially transmitted into the cavity where it undergoes multiple reflections within the cavity.

perimental results that have employed cavities in spin-related studies in quantum dots.

Perhaps the simplest cavity design is a planar Fabry–Pérot resonator [174], shown in Figure 4.8. Such a cavity consists of two parallel partially reflective surfaces. When light is incident on one side, some of the light is transmitted into the region between the surfaces. A portion of this light then reflects back and forth. If the spacing between the reflective surfaces is such that the multiple reflections interfere constructively, then a resonance occurs at this wavelength. When such a condition is satisfied, light builds up inside the cavity. Due to the finite reflectivity of the surfaces, some of the light is also transmitted out the other side. It is relatively straightforward to calculate the transmission coefficient to be

$$T = \frac{1}{1 + F \sin^2 \frac{\delta}{2}} \, , \tag{4.11}$$

where *F* is the "finesse" of the cavity defined in terms of the reflectivity of the surfaces, *R*, as

$$F = \frac{4R}{(1 - R)^2} \, . \tag{4.12}$$

In equation (4.11), δ is the phase shift of the light after each round trip, and is given by

$$\delta = \left(\frac{2\pi}{\lambda} \right) 2nl \cos \theta \, , \tag{4.13}$$

where λ is the vacuum wavelength of the light, *n* is the index of refraction between the surfaces, *l* is the spacing between the surfaces, and θ is the angle of incidence.

Figure 4.9 shows the calculated transmission for a Fabry–Pérot cavity. The transmission spectrum shows a series of sharp peaks that occur when the resonance

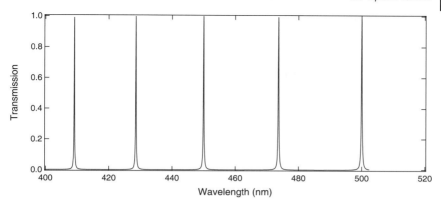

Fig. 4.9 Calculated transmission through a Fabry–Pérot resonator with finesse, $F = 4311$ ($R = 0.97$), $n = 1.5$, $l = 3\,\mu m$, and normal incidence.

condition is satisfied. These peaks are modes of the cavity, and their width decreases with increasing finesse. Another useful quantity for characterizing an optical cavity is the Q-factor, which is defined as the ratio of the wavelength of a cavity mode to the full-width-at-half-maximum of the mode. In Figure 4.9, the cavity modes have $Q \approx 2000$. The fact that there are only certain allowed photon energies within the cavity can be thought of as a concentration of the photon density of states into narrow bands.

If one thinks of the light inside a cavity as bouncing back and forth multiple times before eventually escaping, it is clear that the interaction will be increased between the light and a quantum dot placed inside the cavity. For example, if Faraday rotation is used to measure the spin of electrons in quantum dots embedded in a cavity, the signal will be enhanced with each reflection of the light. The Faraday effect, described in more detail in Chapter 7, yields a spin dependent rotation of the plane of polarization of linearly polarized light. With each pass through the cavity, the light interacts with the spin in the dots and is rotated through an increasingly large angle. The enhancement of Faraday rotation of a quantum dot ensemble by a cavity was investigated in the work of Y. Q. Li *et al.* [175], in which a layer of nanocrystal quantum dots was incorporated within a planar Fabry–Pérot cavity. The reflective surfaces making up the cavity in this case were distributed Bragg reflectors (DBRs), composed of alternating layers of titanium dioxide and silicon dioxide. The differing index of refraction of the two materials results in high reflectivity for light within a specific wavelength range that depends on the layer thicknesses. A first DBR was deposited by electron beam evaporation, capped by a final SiO_2 layer with a width of half the final cavity width. A layer of nanocrystal quantum dots was then chemically deposited on the DBR, and additional layers were evaporated on top to complete the cavity and fabricate a top DBR. As a control, the top DBR was only grown over half of the sample, which allowed comparison of the signal with and without the cavity.

The results are shown in Figure 4.10a. The oscillating signal shows the coherent precession of an ensemble of spins, measured using Faraday rotation. The open symbols show the cavity enhanced signal, and the closed symbols show the data from the same samples, but in a region without the completed cavity. Clearly, a large enhancement of the signal is achieved. The same non-cavity-enhanced data in part 4.10a is shown in Figure 4.10b on a smaller scale. The same spin precession is observed, but with approximately 20 times smaller amplitude. The enhancement is quantified in Figure 4.10c. The enhancement of the signal increases roughly linearly with the cavity Q-factor. This allows the measurement of smaller numbers of quantum dots, or measurement with lower laser intensities.

Another effect of an optical cavity on a quantum dot is due to the change of the photon density of states within the cavity, mentioned above. Interactions between the states in the dot and the light depend on this density of states. For example, the rate of optically driven transitions will depend on this density of states, as can be seen from Fermi's golden rule. This phenomenon is known as the Purcell effect, and will be discussed further in Section 5.5. There has been much work investigating such effects in cavity quantum dot systems, but not too many studies specifically addressing the spin physics. One study that has been performed is described

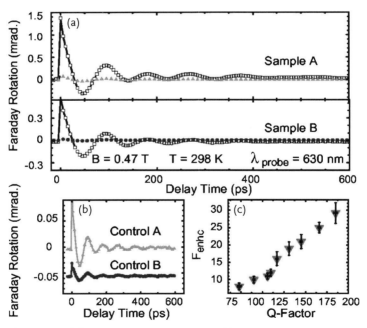

Fig. 4.10 (a) Coherent spin precession measured by cavity enhanced Faraday rotation. Filled symbols show the nonenhanced signal. (b) Nonenhanced data from part (a), shown on a smaller scale. (c) Enhancement of Faraday rotation as a function of cavity Q-factor. Reprinted figure with permission from [175]. Copyright (2006) by the American Institute of Physics.

in S. Ghosh *et al.* [176]. Here, interface fluctuation quantum dots are embedded in microdisk cavities. These cavities consist of a semiconductor disk with a diameter of several microns supported from the bottom by a narrow post. Light can propagate within the disk in a "whispering gallery mode" with the light making successive total internal reflections around the perimeter of the disk. In this work, spins in the quantum dots were measured using time-resolved Kerr rotation (see Chapter 7), and from this, the effective transverse spin lifetime, T_2^*, was extracted. When the cavity was optically pumped sufficiently strongly to produce lasing behavior, an increase in T_2^* was observed. Though the reason for this increase is not entirely understood, it clearly involves the modified interaction between light and the spins in the cavity.

Much of the present work involving quantum dots in cavities is focused on dots embedded in photonic crystal cavities [177]. A photonic crystal can be formed by etching a regular array of holes into a thin semiconductor slab. In the same way that a regular array of atoms in a solid can produce a bandgap for electrons in the material, these holes can produce a bandgap for photons with wavelength on the same order as the hole spacing. By omitting several holes from the center of such a pattern, a small region can be formed without the bandgap present elsewhere. This serves to confine light with certain wavelengths inside this region, forming an optical cavity. If the thickness of the material is chosen appropriately, the light will also be confined in the vertical direction, so the cavity confines light in all three dimensions.

These structures provide very small, high-quality cavities that are readily integrated with self-assembled quantum dot systems. A significant amount of progress has been made in observing coupling between the quantum dot states and the cavity modes (e. g., [178]). However, there has not been much observed to date specifically involving spins in the quantum dots. Nonetheless, there is much promise for spin related studies in these systems. For example, proposals exist to use such integrated cavities for optically-mediated coupling of spins in two quantum dots [179], or fully scalable spin-based quantum computing [180]. Along similar lines, arrays of microdisk cavities may be fabricated on a chip and used to couple multiple quantum dots to perform quantum computing, as was proposed by Imamoglu *et al.* [181] (see Chapter 5 for more discussion of this proposal).

5
Quantum Dots Interacting With the Electromagnetic Field

Quantum dots as described in the previous chapters exhibit interesting types of interaction with the electromagnetic field (which we equally refer to as the radiation field). In this chapter we provide a discussion of radiative transitions between quantum dot states, which involve an exchange of energy with the radiation field. In terms of photons, the energy quanta of the electromagnetic field, these processes describe the absorption and emission of photons by the quantum dot. This is followed by a brief discussion of the driven two-level system, where we introduce the useful concept of the generalized master equation, which takes a dissipative environment into account. We then review the theory of cavity quantum electrodynamics with strong and weak coupling regimes. Finally, we discuss dispersive interaction phenomena of quantum dots and the radiation field, highlighting a few that are relevant to the physics of spins in quantum dots. These include the readout or manipulation of spins via the ac Stark effect and the cavity-mediated coupling of two quantum dots.

5.1
Hamiltonian for Radiative Transitions of Quantum Dots

In this section we discuss electric dipole and magnetic dipole transitions between bound states of a quantum dot. Basically, we review here the usual model for the interaction of an atom with electromagnetic radiation, as discussed in many standard textbooks on quantum optics, for example [182, 183], and apply it to the scenario of a quantum dot.

5.1.1
Electromagnetic Field

The Hamiltonian of the quantized electromagnetic field in the Heisenberg picture is given by

$$H_f(t) = \sum_{\mathbf{k},s} \hbar \omega_k \left(a_{\mathbf{k}s}^{\dagger}(t) a_{\mathbf{k}s}(t) + \frac{1}{2} \right), \tag{5.1}$$

Spins in Optically Active Quantum Dots. Concepts and Methods.
Oliver Gywat, Hubert J. Krenner, and Jesse Berezovsky
Copyright © 2010 WILEY-VCH Verlag GmbH & Co. KGaA, Weinheim
ISBN: 978-3-527-40806-1

where $a_{ks}(t)$ and $a_{ks}^\dagger(t)$ are the photon annihilation and creation operators acting on the mode characterized by the wave vector $\mathbf{k}$ and the polarization s. As already mentioned in the introductory chapter, we do not highlight operators with special notation, assuming that it is clear from the context which symbols are operators. The photon angular frequency ω_k is connected with the wave vector $\mathbf{k}$ by the usual relation $\omega_k = ck$, with $k = |\mathbf{k}|$ and the speed of light c. The term proportional to $1/2$ in Eq. (5.1) is known as the zero-point contribution of the field and is not further discussed here.

5.1.2
Nonrelativistic Electron–Photon Interaction

For the radiative interaction of a quantum dot we first consider the nonrelativistic Hamiltonian of an electron interacting with the electromagnetic field. The action of the electromagnetic field on matter is described in terms of the scalar potential Φ and the vector potential $\mathbf{A}$. For simplicity we set $\Phi = \text{const.}$, here, such that the electric field is generated by the vector potential only,

$$\mathbf{E}(\mathbf{r}, t) = -\nabla \Phi - \partial_t \mathbf{A}(\mathbf{r}, t) = -\partial_t \mathbf{A}(\mathbf{r}, t) , \tag{5.2}$$

where ∂_t is the partial derivative with respect to time t, and ∇ is the nabla operator, with $\nabla \Phi = 0$ in our case. The magnetic field is given by the rotation of the vector potential,

$$\mathbf{B}(\mathbf{r}, t) = \nabla \times \mathbf{A}(\mathbf{r}, t) , \tag{5.3}$$

where $\times$ is the usual vector product. We assume that the electron moves in a time-independent crystal potential $V(\mathbf{r})$ and has the free mass m_0, the electric charge $q = -e$, and the spin operator $\mathbf{S}$. The interaction Hamiltonian for this situation is given by

$$H_{\text{nonrel}} = \frac{1}{2m_0} \left[\mathbf{p} - q\mathbf{A}(\mathbf{r}, t) \right]^2 + V(\mathbf{r}) - \frac{q}{m_0} \mathbf{S} \cdot \mathbf{B}(\mathbf{r}, t) . \tag{5.4}$$

For the vector potential it is convenient to choose the Coulomb (or transverse) gauge, $\nabla \cdot \mathbf{A}(\mathbf{r}, t) = 0$, such that the electron momentum operator $\mathbf{p}$ and the vector potential of the quantized field $\mathbf{A}$ commute. This can be shown using the explicit form of $\mathbf{A}$, Eq. (5.6), below. After applying the Coulomb gauge in Eq. (5.4) we can immediately write

$$H_{\text{nonrel}} = \frac{\mathbf{p}^2}{2m_0} + V(\mathbf{r}) - \frac{q}{m_0} \mathbf{A}(\mathbf{r}, t) \cdot \mathbf{p} + \frac{q^2}{2m_0} \mathbf{A}^2(\mathbf{r}, t)$$
$$- \frac{q}{m_0} \mathbf{S} \cdot \mathbf{B}(\mathbf{r}, t). \tag{5.5}$$

For the electromagnetic field in vacuum, the quantized expression for the vector potential reads

$$
\begin{aligned}
\mathbf{A}(\mathbf{r}, t) &= \sum_{\mathbf{k},s} \sqrt{\frac{\hbar}{2\epsilon \omega_k V_{\text{mode}}}} \left[a_{\mathbf{k}s}(0)\mathbf{e}_{\mathbf{k}s} e^{i(\mathbf{k}\cdot\mathbf{r}-\omega_k t)} + a_{\mathbf{k}s}^{\dagger}(0)\mathbf{e}_{\mathbf{k}s}^{*} e^{-i(\mathbf{k}\cdot\mathbf{r}-\omega_k t)} \right] \\
&= \sum_{\mathbf{k},s} A_k \left[a_{\mathbf{k}s}(0)\mathbf{e}_{\mathbf{k}s} e^{i(\mathbf{k}\cdot\mathbf{r}-\omega_k t)} + a_{\mathbf{k}s}^{\dagger}(0)\mathbf{e}_{\mathbf{k}s}^{*} e^{-i(\mathbf{k}\cdot\mathbf{r}-\omega_k t)} \right].
\end{aligned}
$$
(5.6)

Here, $\mathbf{e}_{\mathbf{k}s}$ is the unit polarization vector for the mode with wave vector $\mathbf{k}$ and polarization s, ϵ is the dielectric constant (in vacuum, $\epsilon = \epsilon_0$), and V_{mode} is the quantization volume of the electromagnetic field. We choose all dimensions of V_{mode} much bigger than the photon wavelength λ, such that the plane wave expansion of $\mathbf{A}$ as shown in Eq. (5.6) exists, which is our scenario of choice in the following. In turn, for photon modes confined by a cavity or resonator, the cavity defines the mode quantization volume. For some cavities with dimensions on the order of a wavelength, a numerical calculation may actually be necessary to accurately obtain the fields present in the cavity modes. Every mode with wave vector $\mathbf{k}$ has two possible polarizations, $s = 1, 2$. The two unit polarization vectors $\mathbf{e}_{\mathbf{k}s}$ are by definition orthonormal, $\mathbf{e}_{\mathbf{k}s}^{*} \cdot \mathbf{e}_{\mathbf{k}'s'} = \delta_{ss'}\delta_{\mathbf{k}\mathbf{k}'}$, where the asterisk $*$ denotes complex conjugation. They are further oriented transversal to the photon propagation direction, $\mathbf{k} \cdot \mathbf{e}_{\mathbf{k}s} = 0$. It is usual to order them such that they form a right-handed system with the photon propagation direction, $\mathbf{e}_{\mathbf{k}1} \times \mathbf{e}_{\mathbf{k}2} = \mathbf{k}/k$. The unit polarization vectors $\mathbf{e}_{\mathbf{k}s}$ are oriented parallel to the electric field vector $\mathbf{E}(\mathbf{r}, t)$, which follows from Eq. (5.2). Obviously, for a linear polarization basis (such as $s = H, V$ for horizontal and vertical polarization) both polarization vectors of a given $\mathbf{k}$ are real-valued. For the circular polarization basis ($s = \sigma^+, \sigma^-$ for right- and left-handed circular polarization) we choose the complex unit polarization vectors $\mathbf{e}_{\mathbf{k}\sigma^\pm} = (1, \pm i)/\sqrt{2}$ in the plane perpendicular to $\mathbf{k}$.

5.1.3
Total Hamiltonian for a Quantum Dot and a Field

For the total Hamiltonian H of the electron interacting with the quantized electromagnetic field we obtain from the above considerations

$$
H = H_{\text{nonrel}} + H_f = H_0 + H_{\text{int}} + H_f ,
$$
(5.7)

where H_f is given by Eq. (5.1), and we have ordered the remaining terms as

$$
H_0 = \frac{\mathbf{p}^2}{2m_0} + V(\mathbf{r}) ,
$$
(5.8)

$$
H_{\text{int}} = -\frac{q}{m_0}\mathbf{A}(\mathbf{r}, t) \cdot \mathbf{p} - \frac{q}{m_0}\mathbf{S} \cdot \mathbf{B}(\mathbf{r}, t) .
$$
(5.9)

Here, we have neglected the term $\propto \mathbf{A}^2(\mathbf{r}, t)$ in Eq. (5.5), which describes photon–photon scattering, assuming that it is small compared to the term linear in $\mathbf{A}$. This is true for low-intensity fields and is a good approximation for most optical experiments.

We now specialize to an electron in a quantum dot interacting with the electromagnetic field. For simplicity we consider a single quantum dot in the crystal ground state and at temperature $T = 0$ K. The potential $V(\mathbf{r})$ is the total crystal potential of the quantum dot. For a quantum dot with a size much smaller than the photon wavelength and that is embedded in a material with dielectric number ϵ_r, the dielectric constant to be chosen in $\mathbf{A}(\mathbf{r}, t)$ is $\epsilon = \epsilon_r \epsilon_0$. For a quantum dot in the strong confinement regime, we can basically apply the standard model of a two-level system interacting with the quantized electromagnetic field from quantum optics. We assume in the following that there are stationary states of the time-independent Hamiltonian H_0 of the quantum dot, of which the two confined single-particle states $|1\rangle$ and $|2\rangle$ are coupled by the radiation field. When discussing the selection rules for electric and magnetic dipole transitions we specify the conditions for the coupling of such states. For this chapter it will be sufficient that there are two stationary states $|1\rangle$ and $|2\rangle$ of H_0 and we need no further information about H_0.

We now transform H into the Schrödinger picture where the time dependence of the field is removed. We obtain for the electron momentum coupled to the vector potential

$$\mathbf{A}(\mathbf{r}) \cdot \mathbf{p} = \sum_{\mathbf{k},s} A_k \left(a_{\mathbf{k}s} e^{i\mathbf{k}\cdot\mathbf{r}} \mathbf{e}_{\mathbf{k}s} \cdot \mathbf{p} + a_{\mathbf{k}s}^\dagger e^{-i\mathbf{k}\cdot\mathbf{r}} \mathbf{e}_{\mathbf{k}s}^* \cdot \mathbf{p} \right). \tag{5.10}$$

For quantum dots, the wavelength of the coupled photon mode is typically much larger than the spatial extension of the confined electron wave functions involved in the transition. Hence, $\mathbf{k} \cdot \mathbf{r} \ll 1$, and we can perform a multipole expansion,

$$e^{\pm i\mathbf{k}\cdot\mathbf{r}} = 1 \pm i\mathbf{k} \cdot \mathbf{r} + \dots, \tag{5.11}$$

which we may truncate after the lowest nonvanishing term in the transition matrix elements taken from Eq. (5.10). In the remainder of this chapter, we discuss the first two interaction terms of the multipole expansion. The first term gives rise to electric dipole transitions, whereas the second term gives rise to magnetic dipole and electric quadrupole transitions.

5.2
Electric Dipole Transitions

At optical wavelengths, that is in the range $\lambda \approx 380$ nm ... 780 nm, most transitions in matter that are induced by electromagnetic radiation are electric dipole transitions [183]. Mathematically, the dipole approximation corresponds to setting $\exp(\pm i\mathbf{k} \cdot \mathbf{r}) \approx 1$ in Eq. (5.10) and inserting the resulting expression into Eq. (5.9). The dipole approximation is exact for a point-like emitter, for which the position of the electron can be chosen as $r = 0$, and therefore the spatial variation of the electromagnetic field is not important, apart from its polarization, which is discussed in

Section 5.2.1. For typical quantum dots, the dipole approximation is usually a good approximation. To give an example, for self-assembled InAs or InGaAs quantum dots the energetically lowest interband transitions are typically in the near infrared ($k \sim 1\,\mu\mathrm{m}^{-1}$), while r is on the order of the quantum dot size (usually a few tens of nanometers). Thus, $kr \sim 10^{-2}$ in this example. Further, the term $e\mathbf{S} \cdot \mathbf{B}(\mathbf{r}, t)/m_0 c$ in Eq. (5.9) can safely be neglected in the dipole approximation because its transition matrix elements are smaller by a factor $kr(\ll 1)$ than the transition matrix elements of $\mathbf{A} \cdot \mathbf{p}$ [183]. Altogether, Eq. (5.9) simplifies in the dipole approximation to

$$H_{\text{int}} \approx -\frac{q}{m_0} \sum_{\mathbf{k},s} A_k \left(a_{\mathbf{k}s} \mathbf{e}_{\mathbf{k}s} \cdot \mathbf{p} + a_{\mathbf{k}s}^{\dagger} \mathbf{e}_{\mathbf{k}s}^{*} \cdot \mathbf{p} \right). \tag{5.12}$$

The scalar products in Eq. (5.12) indicate that the coupling strength of the electron and the electromagnetic field depends on the cosine of the angle between $\mathbf{e}_{\mathbf{k}s}$ (or $\mathbf{e}_{\mathbf{k}s}^{*}$) and $\mathbf{p}$. The coupling is zero if an involved electron momentum is aligned perpendicular to $\mathbf{e}_{\mathbf{k}s}$.

In the following we describe the quantum dot as a two-level system with the ground state $|1\rangle$ and the excited state $|2\rangle$, where $H_0|i\rangle = E_i|i\rangle$ for $i = 1, 2$. For simplicity, we consider the interaction with one single photon mode and drop the summation over $\mathbf{k}$ and s. We assume that the mode is resonant with the transition, $\omega_k = (E_2 - E_1)/\hbar$, or quasi-resonant, $|\omega_k - (E_2 - E_1)/\hbar| \leq \delta_{1,2}$, where $\delta_{1,2}$ is the linewidth of the transition between $|1\rangle$ and $|2\rangle$. For further insight into the two-level dynamics we multiply H_{int} from the left- and the right-hand sides with the unity operator of the two-level system, $|1\rangle\langle1| + |2\rangle\langle2|$. The diagonal momentum matrix elements vanish because $\mathbf{p}$ changes the parity of the state it is applied to. We introduce the raising operator $\sigma_+ = |2\rangle\langle1|$ and the lowering operator $\sigma_- = |1\rangle\langle2|$ and obtain

$$\begin{aligned} H_{\text{int}} \approx -\frac{q}{m_0} A_k \Big[&a_{\mathbf{k}s} \left(\sigma_+ \mathbf{e}_{\mathbf{k}s} \cdot \langle2|\mathbf{p}|1\rangle + \sigma_- \mathbf{e}_{\mathbf{k}s} \cdot \langle1|\mathbf{p}|2\rangle \right) \\ &+ a_{\mathbf{k}s}^{\dagger} \left(\sigma_+ \mathbf{e}_{\mathbf{k}s}^{*} \cdot \langle2|\mathbf{p}|1\rangle + \sigma_- \mathbf{e}_{\mathbf{k}s}^{*} \cdot \langle1|\mathbf{p}|2\rangle \right) \Big]. \end{aligned} \tag{5.13}$$

We note that the Hamiltonian Eq. (5.13) contains energy nonconserving terms, namely the emission of a photon combined with an excitation of the quantum dot, and the Hermitian conjugate expression of this process. However, these terms vanish due to angular momentum selection rules if we consider coupling with a circularly polarized mode (cf. [182], p. 752). For the coupling with a mode of arbitrary polarization, the rotating wave approximation (RWA) may be applied. In the RWA only the energy-conserving interaction terms in Eq. (5.13) are kept. This is usually a good approximation when the mode and the emitter are resonant or quasi-resonant, as we have assumed here, and can be justified as follows: We transform into a frame rotating with the angular frequency ω of the field. In this frame, the energy nonconserving terms rotate at a frequency 2ω and are averaged out to zero on a time scale larger than $\sim 1/2\omega$, which is fast for optical frequencies when compared to typical radiative lifetimes. Assuming a circularly polarized mode, or alternatively, applying the RWA to a mode with noncircular polarization, we obtain the usual interaction Hamiltonian for electric dipole transitions,

$$H_{\text{int}}^{\text{ED}} = -\frac{q}{m_0} A_k \left(a_{\mathbf{ks}} \sigma_+ \mathbf{e_{ks}} \cdot \langle 2|\mathbf{p}|1\rangle + a_{\mathbf{ks}}^\dagger \sigma_- \mathbf{e_{ks}^*} \cdot \langle 1|\mathbf{p}|2\rangle \right). \tag{5.14}$$

With this interaction Hamiltonian we obtain the famous Jaynes–Cummings Hamiltonian [184] from cavity quantum electrodynamics (cavity QED) [185],

$$H = H_0 + H_f + H_{\text{int}}^{\text{ED}}. \tag{5.15}$$

Before we discuss the conditions for nonvanishing electric dipole transition matrix elements, we introduce two frequently encountered quantities that characterize the transition efficiency, namely, the optical Rabi frequency,

$$\hbar \Omega_{\mathbf{ks}} = \left| \frac{q}{m_0} A_k \mathbf{e_{ks}} \cdot \langle 2|\mathbf{p}|1\rangle \right| = \frac{e}{m_0} \sqrt{\frac{\hbar}{2\epsilon_r \epsilon_0 \omega_k}} \, |\mathbf{e_{ks}} \cdot \langle 2|\mathbf{p}|1\rangle|, \tag{5.16}$$

and the oscillator strength,

$$f_{2,1} = \frac{2|\mathbf{e_{ks}} \cdot \langle 2|\mathbf{p}|1\rangle|^2}{m_0 |E_2 - E_1|}. \tag{5.17}$$

The optical Rabi frequency is basically analogous to the usual Rabi frequency of a spin driven by an external ac field (which is typically a magnetic dipole transition, see Section 5.3). The oscillator strength is named after a classical model of oscillators coupled to the electromagnetic field. In this classical model, the oscillator strength describes the fraction of oscillators that are effectively coupled to the field. The above expression for $f_{2,1}$ is a generalization of the classical expression.

5.2.1
Electric Dipole Selection Rules

The scalar product of the unit polarization vector with the momentum matrix element lies at the core of the electric dipole interaction of an electron and a photon. Let us first consider the momentum matrix element $\langle 2|\mathbf{p}|1\rangle$ alone, which is nonzero only if the states $|1\rangle$ and $|2\rangle$ satisfy certain symmetry criteria. These symmetry criteria are called the electric dipole selection rules. As the momentum operator changes the parity of a state (from even to uneven, and vice versa), the momentum matrix element is nonzero only if the states $|1\rangle$ and $|2\rangle$ have different parity. Further criteria are obtained depending on the exact form of the states $|1\rangle$ and $|2\rangle$.

As a first example, we consider orbital angular momentum eigenstates $|1\rangle = |l, m, m_s\rangle$ and $|2\rangle = |l', m', m_s'\rangle$, where l (l'), m (m'), and m_s (m_s') are the quantum numbers of the orbital angular momentum, its projection, and the spin, respectively, which are all defined along an axis, for example z. For these states, the selection rules

$$l - l' = \pm 1, \quad m - m' = 0, \pm 1 \quad \text{and} \quad m_s = m_s' \tag{5.18}$$

must be satisfied [183]. Note that the third of the above conditions, the conservation of the spin orientation, follows directly from the absence of spin operators in the matrix element.

As a second example, we consider the selection rules in the presence of spin–orbit coupling, $H_{so} = \lambda_{so} \mathbf{L} \cdot \mathbf{S}$, as given by Eq. (3.3). In this situation, we apply Clebsch–Gordan theory to couple the spin 1/2 and the orbital angular momentum and switch to the basis of total angular momentum states, $|1\rangle = |j, j_z; l, m, m_s\rangle$ and $|2\rangle = |j', j_z'; l', m', m_s'\rangle$, which diagonalizes Eq. (3.3). The above electric dipole selection rules then transform into

$$j - j' = 0, \pm 1 \,, \quad l - l' = \pm 1 \,, \quad \text{and} \quad j_z - j_z' = 0, \pm 1 \,. \tag{5.19}$$

The electric dipole selection rules stated in Eqs. (5.18) and (5.19) are the usual rules that apply to atomic transitions as well [183]. The transitions with $j_z - j_z' = \pm 1$ are circularly or σ polarized, as they involve net transfer of angular momentum, while the $j_z - j_z' = 0$ transitions are linearly or π polarized.

Finally, the scalar product of the unit polarization vector of the photon and the momentum matrix element of the electron imposes a geometrical condition on the electric dipole interaction. To elaborate further on this property we need to take the crystal symmetry into account for the states $|1\rangle$ and $|2\rangle$, which is studied in more detail in the following.

5.2.2
Interband Transitions in a III–V Semiconductor

Regarding the optical control and detection of spin, we have seen above that the radiative transitions with typically the largest coupling strength – the electric dipole transitions – are actually spin conserving. However, as we recall from the previous chapter, in many types of semiconductors there is spin–orbit coupling present that couples the electron spin with the angular momentum of the Bloch state, for example in the *p*-type valence band states. For many spin-related optical schemes, the spin–orbit coupling is the key to access spins via optical polarization.

To illustrate this, we work out the electric dipole transition matrix elements of a III–V semiconductor quantum dot. According to Eq. (5.19), a transition between the *c* band ($j = 1/2$) and the *v* band ($j = 3/2$) edges satisfies the selection rule for *j*. For simplicity, we consider the coupling with a single photon mode in a Fock state $|n_{\mathbf{ks}}\rangle$ where the photon population number is given by $n_{\mathbf{ks}}$. We take into account the envelope function approximation from Chapter 4 and consider the states

$$|1\rangle = |\phi_c\rangle |u_{J_z}^c\rangle |n_{\mathbf{ks}} + 1\rangle \quad \text{and} \quad |2\rangle = |\phi_v\rangle |u_{J_z'}^v\rangle |n_{\mathbf{ks}}\rangle \,, \tag{5.20}$$

where $|\phi_b\rangle$ is the envelope function and $|u_{J_z}^b\rangle$ the Bloch function of band $b = c, v$ with angular momentum projection J_z. For the modulus of the photon absorption matrix element, for example, we obtain after a calculation that we leave as an exercise to the reader,

$$|\langle 2|H_{\text{int}}^{\text{ED}}|1\rangle| = \frac{e}{m_0} A_{\mathbf{k}} \sqrt{n_{\mathbf{k}s}+1} \left| \mathbf{e}_{\mathbf{k}s} \cdot \langle u_{J_z}^c |\mathbf{p}| u_{J_z'}^v \rangle \int \mathrm{d}^3 r \phi_c(\mathbf{r},\sigma) \phi_v^*(\mathbf{r},\sigma) \right|.$$

(5.21)

As we can see in the above equation, it turns out that the electric dipole transition matrix element between the c and v bands is a product of an overlap integral of the corresponding envelope functions and a momentum matrix element of the c and v Bloch states of the semiconductor crystal.

The overlap integral of the envelope functions determines between which confined levels an optical transition is allowed. Obviously, for the harmonic oscillator states mentioned earlier in Section 2.3.2 recombination will only occur between electron and hole states with the same orbital symmetry.

The symmetry properties of the Bloch functions Eqs. (3.6–3.10) provide further insights into the structure of the momentum matrix element. The zincblende lattice has cubic symmetry, thus the matrix elements $\langle s|p_\alpha|\alpha\rangle = p_{cv}$ are equivalent for $\alpha = x, y, z$. This interband momentum matrix element p_{cv} is a material constant that is related to the Kane energy $E_p = 2p_{cv}^2/m_0$, which can be found in tables, for example, Appendix C.

We now take the scalar product with the unit polarization vector into account and consider a photon that propagates in a direction given by the polar angles (θ, ϕ) with respect to a coordinate system coinciding with the cubic crystal axes. When we calculate the momentum matrix elements with the Bloch functions we note that there are circularly polarized transitions from the hh band as well as the lh band to the c band. For the circularly polarized hh–c transition, the transition matrix element is obtained up to a global phase as

$$\mathbf{e}_{\mathbf{k}s} \cdot \langle u_{\sigma/2}^c |\mathbf{p}| u_{\sigma 3/2}^v \rangle = \frac{p_{cv}}{2} e^{i\sigma\phi} (\cos\theta - \sigma s).$$

(5.22)

Here, $\sigma = \pm 1$ is a parameter related to the spin $\sigma/2$ of the electron involved, which also assures spin conservation, and $s = \pm 1$ is the photon circular polarization $\sigma^{\pm}$. For the lh transition, in turn, we obtain up to a global phase

$$\mathbf{e}_{\mathbf{k}s} \cdot \langle u_{\sigma/2}^c |\mathbf{p}| u_{-\sigma/2}^v \rangle = \frac{p_{cv}}{2\sqrt{3}} e^{-i\sigma\phi} (\cos\theta - \sigma s).$$

(5.23)

The different prefactors of above two expressions Eqs. (5.22) and (5.23) are responsible for the famous ratio 1/3 of the optical transition rates for lh and hh states. For lh, there is also a linearly polarized transition to the c band due to the $|z\rangle$ part of the Bloch states. This transition matrix element is up to a global phase given by

$$\mathbf{e}_{\mathbf{k}s} \cdot \langle u_{\sigma/2}^c |\mathbf{p}| u_{\sigma/2}^v \rangle = \frac{p_{cv}}{\sqrt{3}} \sin\theta.$$

(5.24)

For the case of spontaneous emission, the above matrix elements suggest that the polarization of the emitted photon depends not only on the spin of the recombining

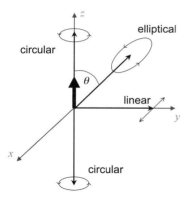

Fig. 5.1 Illustration of the different photon polarizations from a given σ transition upon emission into different directions. The central arrow indicates a dipole transition moment along z, the thin black arrows are photon propagation directions, and the arrows near their endings show the corresponding polarizations.

exciton, but also on the direction in which the photon is detected after emission. To illustrate this with an example, for a hh transition with angular momentum $+1$ transferred to the field, and photon emission along the angular momentum quantization axis, say z, the photon is of circular polarization σ^+. For emission at an angle θ with respect to z, the photon polarization can be viewed as the result of a projection of the angular momentum onto the photon propagation direction. It is given by a superposition of the two photon polarizations $\sigma^\pm$ with relative amplitudes according to Eq. (5.22). For an increasing θ, the photon polarization thus evolves from circular to elliptical, as an increasing component of the opposite circular polarization is admixed. For emission exactly perpendicular to z, the photon polarization is linear. If θ increases even further, the photon polarization is elliptical again, until for $\theta = \pi$ the photon polarization is σ^-, the circular polarization opposite to the one for $\theta = 0$. The different photon polarizations for different emission directions are illustrated in Figure 5.1.

We conclude the section on electric dipole transitions with two comments related to the classical description.

5.2.3
Equivalent Classical Electric Dipole Picture

In the dipole approximation actually only the electric field is involved in the interaction, and the magnetic field **B** is neglected completely. Especially, the interaction of **B** with the spin is negligible because it is proportional to $\mathbf{k} \cdot \mathbf{r} \ll 1$, which is on the order of the terms that are dumped in the dipole approximation. Further, the orbital effect of **B** due to the Lorentz force is also irrelevant in the dipole approximation, which seems clear in view of the fact that the coupling term $\mathbf{A} \cdot \mathbf{p}$ can in the dipole approximation be transformed right away into a coupling of the form

$\mathbf{d} \cdot \mathbf{E}$, where $\mathbf{d} = e\mathbf{r}$ is the electric dipole moment operator and $\mathbf{E}$ the electric field. To see this we consider the classical Hamiltonian for the interaction of the electric dipole moment $\mathbf{d}$ with the electric field, $H_{\text{classic}}^{\text{ED}} = -\mathbf{d} \cdot \mathbf{E}$, and insert the vector potential of the quantized field using the definition $\mathbf{E} = -\partial_t \mathbf{A}$. If we use a similar representation in terms of the states $|1\rangle$ and $|2\rangle$ as for H_{int}, we obtain

$$H_{\text{classic}}^{\text{ED}} = -i\omega_k A_{\mathbf{k}} \left[a_{\mathbf{k}s} \left(\sigma_+ \mathbf{e}_{\mathbf{k}s} \cdot \langle 2|\mathbf{d}|1\rangle + \sigma_- \mathbf{e}_{\mathbf{k}s} \cdot \langle 1|\mathbf{d}|2\rangle \right) \right.$$
$$\left. - a_{\mathbf{k}s}^{\dagger} \left(\sigma_+ \mathbf{e}_{\mathbf{k}s}^* \cdot \langle 2|\mathbf{d}|1\rangle + \sigma_- \mathbf{e}_{\mathbf{k}s}^* \cdot \langle 1|\mathbf{d}|2\rangle \right) \right]. \tag{5.25}$$

We now go back to our quantum mechanical interaction Hamiltonian Eq. (5.13) in the dipole approximation. We notice that the momentum operator can by replaced by the commutator of $\mathbf{r}$ and H_0 by using the relation $[\mathbf{r}, H_0] = i\hbar\mathbf{p}/m_0$. For $\omega_k = (E_2 - E_1)/\hbar$ we can replace all momentum matrix elements by expressions of the form

$$\langle 2|\mathbf{p}|1\rangle = \frac{im_0}{\hbar}\langle 2|[H_0, \mathbf{r}]|1\rangle = \frac{im_0}{\hbar}(E_2 - E_1)\langle 2|\mathbf{r}|1\rangle$$
$$= im_0\omega_k\langle 2|\mathbf{r}|1\rangle . \tag{5.26}$$

It is an easy exercise to verify that we thus obtain the Hamiltonian Eq. (5.25) from H_{int} in the electric dipole approximation. This classical correspondence leads us to the concept of an electric dipole moment of a transition. This in some sense provides a more intuitive picture of an electric dipole transition than the momentum matrix element. The electric field of a photon can induce a dipole transition only if a "jump" of the atomic wave function (or of the Bloch states in a semiconductor) that provides a suitable transition dipole moment can be accomplished.

5.2.4
Semiclassical Interaction with a Laser Field

For a coherent photon state $|\alpha\rangle$, the semi-classical interaction Hamiltonian of a two-level system coupled to a classical field is recovered from the full quantum mechanical expression Eq. (5.14). A coherent photon state $|\alpha\rangle$ is defined by

$$|\alpha\rangle = \exp\left(\alpha a_L^{\dagger} - \alpha^* a_L\right)|0\rangle , \tag{5.27}$$

where $\alpha \in \mathbb{C}$, $a_L^{(\dagger)}$ are photon operators of the laser mode, $|0\rangle$ is the photon vacuum, and the mean number of photons in the mode is given by $|\alpha|^2$. Since coherent states are eigenstates of the annihilation operator,

$$a_L|\alpha\rangle = \alpha|\alpha\rangle , \tag{5.28}$$

the photon operators in H_{int} are simply replaced by complex numbers α and α^*, respectively. Provided that the interaction with the two-level system does not significantly change the state of a laser field, a coherent state is obviously a good approximation for the laser field. Hence, a semiclassical interaction description is usually sufficient for the interaction of a laser driving a two-level system [182].

5.3
Magnetic Dipole Transitions

The second term in the multipole expansion may be the dominant term in a transition matrix element if the electric dipole term vanishes due to selection rules. This second term ($\propto \mathbf{k} \cdot \mathbf{r}$) leads to momentum matrix elements of the form $\langle i|(\mathbf{k} \cdot \mathbf{r})(\mathbf{e}_{ks} \cdot \mathbf{p})|j\rangle$ in H_{int}. We add to this term the interaction $\propto \mathbf{S} \cdot \mathbf{B}$ of the spin with the magnetic field. The resulting interaction term can be represented as the sum of a magnetic dipole term and an electric quadrupole term [183]. The magnetic dipole term describes the coupling of the total magnetic dipole moment of the electron to the magnetic field. For transitions between angular momentum eigenstates $|l, m_l, m_s\rangle$ and $|l', m_{l'}, m_{s'}\rangle$ with orbital and spin magnetic moment m_l and m_s, respectively, the selection rules for magnetic dipole transitions are

$$l - l' = 0 , \quad m_l - m_{l'} = 0, \pm 1 , \quad \text{and} \quad m_s - m_{s'} = 0, \pm 1 . \tag{5.29}$$

In the presence of the spin–orbit interaction $H_{so} = \lambda \mathbf{L} \cdot \mathbf{S}$, as in Section 5.2, the magnetic dipole selection rules transform into

$$l - l' = 0 , \quad j - j' = 0, \pm 1 , \quad \text{and} \quad m_j - m_{j'} = 0, \pm 1 . \tag{5.30}$$

The selection rules for electric quadrupole transitions are given by $l - l' = 0, \pm 2$ and $m_l - m_{l'} = 0, \pm 1, \pm 2$.

The electric quadrupole transition can be interpreted as the interaction of the electric quadrupole moment with the gradient of the electric field [183]. It is usually experimentally possible to place the emitter at a location, for example, inside a cavity, where the magnetic field is large and the gradient of the electric field is negligible, enabling the excitation of only a magnetic dipole transition. Typical examples for magnetic dipole transitions are magnetic resonance transitions, such as electron spin resonance.

5.4
Generalized Master Equation of the Driven Two-Level System

We provide a brief overview of the dynamics of two levels that are coupled by an external oscillating field. Additionally, the environment is coupled to the two-level system, which induces transitions and affects the coherent time evolution of the two-level system. In this section we present the generalized master equation description of this problem. It is a straightforward task to apply the master equation formalism to a simple two-level system, and we find it useful to work out a few details here.

5.4.1
The Driven Two-Level System

To be specific, we consider a spin $1/2$ in a static magnetic field $\mathbf{B}_z$ along z with the ground state $|\uparrow\rangle$ and the excited state $|\downarrow\rangle$. The spin is driven by a circularly polarized magnetic field $\mathbf{B}_\perp(t)$, which rotates in the xy plane with frequency ω. The Hamiltonian of this system is well known from electron spin resonance [1] and can be written as

$$H_S = \frac{\hbar\Delta}{2}\sigma_z + \frac{\hbar\Omega}{2}\left[\cos(\omega t)\sigma_x + \sin(\omega t)\sigma_y\right], \tag{5.31}$$

where $\Delta = g_e\mu_B B_z$ and $\Omega = g_e\mu_B B_\perp$. Here we have assumed an isotropic g factor of the electron, g_e, which applies to c band electrons in many semiconductors. For a linearly polarized magnetic field we would obtain the same Hamiltonian after decomposing the field into two counter-rotating components and keeping only one of them after applying the RWA. It is important to note that in the linearly polarized case only half the amplitude of the magnetic field enters in Ω.

We now rewrite the above Hamiltonian in terms of the density operator ρ of the spin $1/2$, which consists of the matrix elements $\rho_{\uparrow\uparrow}$, $\rho_{\downarrow\downarrow}$, $\rho_{\downarrow\uparrow}$, and $\rho_{\uparrow\downarrow}$, where we apply the notation $\rho_{nm} = |n\rangle\langle m|$. The off-diagonal matrix elements are complex conjugates, $\rho_{\uparrow\downarrow} = \rho_{\downarrow\uparrow}^*$, since ρ is Hermitian. Using ρ_{nm} instead of the Pauli matrices, the Hamiltonian reads

$$H_S = \frac{\hbar\Delta}{2}\left(\rho_{\downarrow\downarrow} - \rho_{\uparrow\uparrow}\right) + \frac{\hbar\Omega}{2}\left(e^{-i\omega t}\rho_{\uparrow\downarrow} + e^{i\omega t}\rho_{\downarrow\uparrow}\right), \tag{5.32}$$

We now transform into the frame rotating at frequency ω. This is achieved by the unitary operator

$$U = e^{-\frac{i\omega t}{2}\left(\rho_{\downarrow\downarrow} - \rho_{\uparrow\uparrow}\right)}, \tag{5.33}$$

which, when applied in the Schrödinger equation, leads us to the transformed Hamiltonian

$$H_S' = U^\dagger H_S U - i U^\dagger \partial_t U$$
$$= \frac{\hbar\delta}{2}\left(\rho_{\downarrow\downarrow} - \rho_{\uparrow\uparrow}\right) + \frac{\hbar\Omega}{2}\left(\rho_{\uparrow\downarrow} + \rho_{\downarrow\uparrow}\right), \tag{5.34}$$

where $\delta = \Delta - \omega$ is the detuning of the field with respect to the level splitting.

5.4.2
System-Reservoir Approach

Having obtained the Hamiltonian H_S (or H_S') of the driven two-level system we can obtain the dynamics of the system by solving the Von Neumann equation. However, we additionally want to take into account the coupling to a reservoir that damps the coherent time evolution of the system [186, 187]. Such a reservoir can be the electromagnetic field with all its modes, for example, or the reservoir of lattice vibrations, the phonons, or any other bath coupling to the two-level system.

For this it is convenient to consider the density operator ρ_{S+R} of the system S and the reservoir R. The total Hamiltonian consists of three parts, $H = H_S + H_R + H_{S-R}$, where H_S describes the coherent dynamics of S, H_R is the Hamiltonian of R, and H_{S-R} is the interaction of S and R which is the term responsible for the damping of S. The density operator of the system, which we have labeled ρ, is obtained by taking the trace over the R degrees of freedom in ρ_{S+R}. The trace over R is now also taken in the equation of motion. In the Born and Markov approximations, the equation of motion for ρ is obtained as a generalized master equation in the Lindblad form [187],

$$\dot{\rho} = \mathcal{L}\rho . \tag{5.35}$$

Here, $\mathcal{L}$ is a superoperator acting on the density operator. For the individual matrix elements, the above equation reads explicitly

$$\dot{\rho}_{nn} = -i\langle n|[H,\rho]|n\rangle + \sum_j W_{nj}\rho_{jj} - \sum_j W_{jn}\rho_{nn} \tag{5.36}$$

$$\dot{\rho}_{nm} = -i\langle n|[H,\rho]|m\rangle - \left[\frac{1}{2}\sum_j (W_{jm} + W_{jn}) + V_m + V_n\right]\rho_{nm} . \tag{5.37}$$

Here, W_{nm} is the relaxation rate for the process leading from state m to state n, and V_n and V_m are called pure decoherence rates.

Consequently, the generalized Master equation of the driven two-level system reads

$$\dot{\rho}_{\uparrow\uparrow} = \frac{\Omega}{2}\left(\rho_{\downarrow\uparrow} - \rho_{\uparrow\downarrow}\right) + W_{\uparrow\downarrow}\rho_{\downarrow\downarrow} - W_{\downarrow\uparrow}\rho_{\uparrow\uparrow} \tag{5.38}$$

$$\dot{\rho}_{\downarrow\downarrow} = \frac{\Omega}{2}\left(\rho_{\uparrow\downarrow} - \rho_{\downarrow\uparrow}\right) + W_{\downarrow\uparrow}\rho_{\uparrow\uparrow} - W_{\uparrow\downarrow}\rho_{\downarrow\downarrow} \tag{5.39}$$

$$\dot{\rho}_{\downarrow\uparrow} = -\left(i\delta + \frac{1}{T_2}\right)\rho_{\downarrow\uparrow} + i\frac{\Omega}{2}\left(\rho_{\downarrow\downarrow} - \rho_{\uparrow\uparrow}\right) . \tag{5.40}$$

Here, $W_{\uparrow\downarrow}$ and $W_{\downarrow\uparrow}$ are the rates of incoherent spin flips that are induced by the interaction with the reservoir. The spin relaxation time T_1 is connected to these rates by $1/T_1 = W_{\uparrow\downarrow} + W_{\downarrow\uparrow}$. Further, in the equation for the off-diagonal matrix element we have introduced the abbreviation

$$\frac{1}{T_2} = \frac{1}{2}\left(W_{\uparrow\downarrow} + W_{\downarrow\uparrow}\right) + V_\uparrow + V_\downarrow . \tag{5.41}$$

Obviously, T_2 is the characteristic decay time of $\rho_{\downarrow\uparrow}$, which identifies it as the decoherence time of the two spin states. We notice that we can rewrite the above equation using T_1,

$$\frac{1}{T_2} = \frac{1}{2T_1} + V_\uparrow + V_\downarrow . \tag{5.42}$$

This indicates that T_2 has a lower bound, $T_2 \leq 2T_1$. We notice that increasing incoherent spin-flip rates increase the decoherence rate as well, adding up to the intrinsic spin decoherence rate, $V_\uparrow + V_\downarrow$. This increase of the decoherence rate due to relaxation processes is of course also true in the more general case for the decoherence of two levels n and m in an arbitrary multilevel system, as we can see in the second term on the right-hand side of Eq. (5.37). Any incoherent process that takes the system out of the states n or m with a certain rate will increase the decay rate of the coherence. Therefore small decoherence rates are only found in strongly isolated systems.

As a special approximation, the adiabatic approximation consists in setting the off-diagonal matrix elements constant, $\dot{\rho}_{\uparrow\downarrow} = 0$. This allows for an algebraic replacement of the off-diagonal matrix elements in Eqs. (5.38) and (5.39). It turns out that the spin populations in the adiabatic approximation are determined by the equations

$$\dot{\rho}_{\uparrow\uparrow} = -(W_{\text{eff}} + W_{\downarrow\uparrow})\rho_{\uparrow\uparrow} + (W_{\text{eff}} + W_{\uparrow\downarrow})\rho_{\downarrow\downarrow} \tag{5.43}$$

$$\dot{\rho}_{\downarrow\downarrow} = (W_{\text{eff}} + W_{\downarrow\uparrow})\rho_{\uparrow\uparrow} - (W_{\text{eff}} + W_{\uparrow\downarrow})\rho_{\downarrow\downarrow} . \tag{5.44}$$

where we have introduced an effective driving rate

$$W_{\text{eff}} = \frac{T_2 \Omega^2}{2(T_2^2 \delta^2 - 1)} , \tag{5.45}$$

which is basically the spin-flip rate induced by the external driving by a field with detuning δ and Rabi frequency Ω. It is easily verified that this rate is of Lorentzian shape as a function of δ, with a width of twice the spin decoherence rate, $2/T_2$. This characteristic quenching of the driving as a function of the detuning δ enables the measurement of the spin decoherence rate, or rather, a lower bound for it (if additional line broadening mechanisms are present), via the transition linewidth in electron spin resonance (ESR). It is clear from the above equations that the dynamics in the adiabatic approximation is dominated by the total spin driving rates

$$W_{\uparrow\downarrow}^{\text{total}} = W_{\text{eff}} + W_{\uparrow\downarrow} \quad \text{and} \quad W_{\downarrow\uparrow}^{\text{total}} = W_{\text{eff}} + W_{\downarrow\uparrow} . \tag{5.46}$$

For the stationary solution, the entire density operator is constant, $\dot{\rho} = 0$ and thus we set $\dot{\rho}_{\uparrow\uparrow} = \dot{\rho}_{\downarrow\downarrow} = 0$ in the above equations. This leads to the following analytical solution for the stationary spin populations, which we denote by $\bar{\rho}_{\uparrow\uparrow}$ and $\bar{\rho}_{\downarrow\downarrow}$,

$$\bar{\rho}_{\uparrow\uparrow} = \frac{W_{\uparrow\downarrow}^{\text{total}}}{W_{\uparrow\downarrow}^{\text{total}} + W_{\downarrow\uparrow}^{\text{total}}} , \tag{5.47}$$

$$\bar{\rho}_{\downarrow\downarrow} = \frac{W_{\downarrow\uparrow}^{\text{total}}}{W_{\uparrow\downarrow}^{\text{total}} + W_{\downarrow\uparrow}^{\text{total}}} , \tag{5.48}$$

which in the case $W_{\uparrow\downarrow} = W_{\downarrow\uparrow}$ are of course both equal to 1/2 due to normalization.

In the frame of a system-reservoir approach as discussed above, we could now also work with a generalized master equation to look at dissipative regimes in the optical interaction, similarly as for electron spin resonance.

The two-level model discussed above is also easily expandable to systems consisting of more than just two levels. There is a big variety of situations in quantum dots for which a generalized master equation description is suitable. An example of this is optically detected resonance of a single electron spin in a quantum dot. Here, we consider the above two spin states plus the two X^{1-} charged exciton orbital ground states with opposite hh spin orientations [188, 189]. This four-level system is driven by a field that induces electron spin resonance, as in the example shown above, and by an additional circularly polarized laser field that drives one optical transition. In a suitable regime of field intensities, the electron spin dynamics modulates the photoluminescence [188], or, alternatively, the photocurrent extracted from the dot [189], which allows one to measure single-electron spin decoherence optically. While this is a first-order optical method to detect spin properties in a quantum dot, there are other fascinating approaches for spin detection and manipulation by second-order interactions with light. These will be discussed in the Sections 5.6, 7.2, and 7.3.8.

Before we continue this chapter with quantum dots interacting with cavity resonators, a few further remarks on electron spin decoherence in quantum dots are due. As summarized, for example, in the tutorial by Cerletti *et al.* [11], the spin–orbit interaction transfers orbital fluctuations affecting an electron, due to phonons for example, into its spin domain; this has been studied for quantum dots by a number of authors in great detail. It has been shown that in leading order these fluctuations are transverse to the direction of an applied magnetic field [190]. As these fluctuations do not contribute to the intrinsic decoherence rate of the spin, $T_2 = 2T_1$ is obtained for this particular decoherence mechanism. Yet, there are usually further decoherence mechanisms for an electron spin in a quantum dot, leading to the often encountered situation where $T_2 \ll 2T_1$. These include the dipolar coupling to other spins and the hyperfine coupling to nuclear spins, which will be discussed in Chapter 6.

Now back again to the interaction with the electromagnetic field. The intensity of the electromagnetic field can be increased in experiments by means of a cavity. In the following we summarize the basics of cavity quantum electrodynamics.

5.5
Cavity Quantum Electrodynamics

We now consider an optically active structure, such as a quantum dot, coupled to a cavity mode, and discuss a few elementary regimes of cavity quantum electrodynamics (cavity QED). We refer to the quantum dot as an emitter in the following. A cavity basically defines boundary conditions for a volume that confines the electromagnetic field. Different kinds of cavities were already introduced in Section 4.2. As already mentioned there, a cavity is usually tailor-made to modify the interaction of a certain emitter with the electromagnetic field in a specific way. If a cavity provides an increased density of photon modes to which an emitter can couple, then the optical interaction can be enhanced. In contrast, if an emitter couples to a frequency range where the cavity provides a smaller mode density than free space, then the interaction strength can also be reduced.

In experiments, the cavity enhancement effect can be maximized basically by three types of adjustments. First, the cavity must be designed such that the emission line of the emitter and the cavity mode match spectrally. Second, the emitter must be located at a position where it couples to the field maximum. Third, the polarization of the emitter's transition must match the mode polarization.

If a cavity mode is the dominant mode to which the emitter couples, then we invoke the Jaynes–Cummings Hamiltonian described in Section 5.2. For the electromagnetic mode volume, the actual cavity mode volume is inserted into the vector potential. Consequently, the amplitude of the cavity electric and magnetic fields increase with decreasing mode volume. The interacting emitter and cavity mode are now united into a joint system. By introducing incoherent transition rates for this system, in the emitter subsystem we can now take into account the coupling of the emitter with a reservoir, for example all other modes of the electromagnetic field. If we assume that this electromagnetic reservoir is in its vacuum state, then we may neglect excitations by it and restrict ourselves to the decay rate W_{em}, which is also called the spontaneous emission rate. The other subsystem, the cavity mode, also exhibits a photon decay rate, W_{cav}, namely the rate at which photons are lost in the cavity, for example by transmission through a not completely reflecting cavity boundary. This decay rate gives rise to a finite linewidth of the cavity mode, $\Delta\omega = W_{cav}$, which is often also expressed by the quality factor $Q = \omega/\Delta\omega = \omega/W_{cav}$. Similarly as for the emitter, if the surrounding electromagnetic field is in the vacuum state, we may neglect processes that feed photons into the cavity mode from the outside.

There are various treatments of a damped emitter in a damped cavity in the literature, for example in the presence of a laser that drives the system [186], which are recommended for further study of such problems. We now focus on the distinction of two important regimes for the Jaynes–Cummings Hamiltonian, the regime of strong coupling and the regime of weak coupling.

5.5.1
Strong Coupling Regime

The strong coupling regime of cavity QED is reached if the coupling of the emitter and the cavity, the Rabi frequency Ω, is by far the largest transition rate in the system. Again, we consider an emitter that is resonant or quasi-resonant with a cavity mode. We diagonalize the Jaynes–Cummings Hamiltonian and obtain eigenstates that are coherent superpositions of emitter and photon states. In fact, if the emitter is prepared in an excited state at a certain point in time, this single energy quantum oscillates coherently between the emitter and the resonant cavity mode at a rate determined by Ω, as is typical for coherent coupling. In a realistic system, decoherence is present, which damps this coherent oscillation.

In the following, we briefly review the theory of the strong coupling regime. For the Hilbert space of the emitter coupled to the cavity mode we use the basis vectors $|i, n\rangle = |i\rangle \otimes |n\rangle$. Here, $|i\rangle = |g\rangle, |e\rangle$ describes the ground and excited states of the emitter with energies E_g and E_e, respectively, and $|n\rangle$ is a photon Fock state with the number of photons n. We switch here to the e and g notation for the emitter for an easier distinction between photon number and emitter states. Clearly, the state $|g, 0\rangle$, consisting of the photon vacuum and the emitter in its ground state, forms an invariant subspace, as there is no other state with the same energy.

After diagonalization of the Jaynes–Cummings Hamiltonian we obtain the well known "dressed states",

$$| + n\rangle = \cos \theta_n |g, n\rangle - i \sin \theta_n |e, n - 1\rangle \tag{5.49}$$

$$| - n\rangle = \sin \theta_n |g, n\rangle + i \cos \theta_n |e, n - 1\rangle \tag{5.50}$$

for $n > 0$, where the mixing angle θ_n characterizes the coupling strength in the presence of n energy quanta versus the detuning and is given by the relation

$$\tan(2\theta_n) = \frac{\Omega \sqrt{n}}{\delta} . \tag{5.51}$$

Here, δ is again the detuning of the field with respect to the transition, $\delta = \Delta - \omega$, where $\hbar\Delta = E_e - E_g$. The eigenstate for zero energy quanta ($n = 0$), as already mentioned above, is $|g, 0\rangle$. The corresponding eigenenergy is given by $E_0 = 0$. For $n > 0$, we obtain

$$E_{\pm n} = \hbar\omega n - \frac{\hbar\delta}{2} \pm \frac{\hbar}{2} \sqrt{n\Omega^2 + \delta^2} . \tag{5.52}$$

The square root in the above term provides the characteristic level repulsion of a strongly coupled system. For $\delta = 0$, the splitting $E_{+n} - E_{-n}$ is just the photon number dependent Rabi splitting. Strong coupling has been achieved for a single quantum dot in a micropillar cavity [191], in a photonic crystal nanocavity [192, 193], and in a microdisk cavity [194].

The strong coupling regime of cavity QED is particularly interesting for quantum information schemes as it allows one to convert localized qubits into flying qubits suitable for transmission of quantum information [195]. If several qubits interact with one cavity mode, then the cavity can act as a bus for quantum information that couples qubits with comparably large spatial separations [196, 197].

5.5.2
Weak Coupling Regime

The weak coupling regime of cavity QED is basically characterized by a Rabi frequency that is smaller than or comparable to other transition rates of the system. In view of the dynamics, we observe in the weak coupling regime exponential decays for the spontaneous photon emission, in contrast to the strong coupling regime, where typically damped oscillations occur. So the weak coupling regime can be defined by a Rabi frequency that is small compared to the optical decoherence time. The most prominent feature of the weak coupling regime is the cavity-induced modification of the photon density of states. As the rates of photon absorption and spontaneous emission according to Fermi's golden rule depend on the energetically available photon modes, an enhancement or reduction of the mode density at the transition frequency will increase or reduce these optical transition rates, respectively. The effect of the cavity-enhanced transition rate has first been described by Purcell for nuclear spin transitions in a radio-frequency cavity and has since been called the Purcell effect [198]. For a single quantum dot, Purcell factors in the range of $\sim 2 \ldots 10$ were obtained already a few years ago in microdisk cavities [199] and in micropillar cavities [200–202]. In addition to the enhancement of spontaneous emission, a suppression of the emission rate by an order of magnitude has also been demonstrated for an off-resonant microcavity in the weak coupling regime [203].

5.6
Dispersive Interaction

For spins in quantum dots, an important class of interaction phenomena is given by the dispersive interaction with the radiation field. Here we especially consider a few types of coherent two-photon processes, which are typically induced by the field of a cavity or by shining high-intensity laser light on a quantum dot. In some of these processes there is a photon being absorbed and a photon of equal energy being emitted simultaneously, and there is no net energy transfer between the quantum dot and the field. Such processes are, for example, also responsible for the refraction of light in matter. The usage of dispersive phenomena goes even further than that, as we are exploring in the following. From the availability of a given second-order transition or, conversely, its blocking due to the Pauli principle, and the back-action on the photon field one can, for example, determine the state of an emitter, thus allowing for optical readout of quantum dot states. Taking advantage of optical selection rules, the effect of Faraday rotation or also Kerr rotation this way even enable the dispersive readout of spin states, as we discuss in Section 7.3.4. Here we first discuss a few fundamental properties of the dispersive regime. A useful technique in this context is the Schrieffer–Wolff transformation [118] as we show in the following.

5.6.1
Lamb Shift and AC Stark Shift

Here we stick with the notation introduced in Section 5.5 and start with the Jaynes–Cummings Hamiltonian of an emitter labeled with i that is coupled to a cavity mode,

$$H = \hbar \Delta_i s_{z,i} + \hbar \Omega_i \left(a s_i^+ + a^\dagger s_i^- \right) + \hbar \omega a^\dagger a \ . \tag{5.53}$$

Here we apply the pseudo-spin operators $s_{z,i} = (1/2)(|e\rangle_i \langle e|_i - |g\rangle_i \langle g|_i)$, $s_i^+ = |e\rangle_i \langle g|_i$ and $s_i^- = |g\rangle_i \langle e|_i$, and the notation $\hbar \Omega_i$ for the interaction energy. We consider the dispersive regime, where the cavity mode is nonresonant with all transitions of the emitter, and the coupling of cavity and emitter in Eq. (5.53) can be treated perturbatively. The coupling of the emitter i and the cavity can be integrated out to leading order by a Schrieffer–Wolff transformation U [181, 197]. Explicitly, we transform to the Hamiltonian

$$H' = U H U^\dagger \ , \tag{5.54}$$

where $U = \exp A_i$ is a unitary operator generated by

$$A_i = \frac{\Omega_i}{\delta_i} \left(a s_i^+ - a^\dagger s_i^- \right) \ . \tag{5.55}$$

As a result we obtain

$$
\begin{aligned}
H' &= e^{A_i} H e^{-A_i} \\
&= H + [A_i, H] + \frac{1}{2}[A_i, [A_i, H]] + \dots \\
&\approx \frac{\hbar \Delta_i}{2} s_{z,i} + \hbar \omega \left(a^\dagger a + \frac{1}{2} \right) + \frac{\hbar \Omega_i^2}{\delta_i} \left(a^\dagger a + \frac{1}{2} \right) s_{z,i} \ .
\end{aligned}
\tag{5.56}
$$

The last line of the above equation is obtained to second order in Ω for the case when $\Omega \sqrt{n} \ll \delta$. The last term in the last line describes the ac Stark shift (being proportional to the photon number operator, $a^\dagger a$) and the Lamb shift (being proportional to $1/2$).

In the above equation we can see that, depending on the state of the qubit, the cavity resonance shifts by $\pm \Omega_i^2/\delta_i$ due to the ac Stark shift, where δ_i is the photon detuning with respect to the transition of the emitter i. A measurement of the cavity resonance shift hence provides a readout mechanism for the qubit. This has beautifully been demonstrated for a Cooper pair box at the charge degeneracy point coupled to a stripline resonator [204], which is a system with excellent conditions to realize quantum optics schemes in a solid state environment. In turn, the ac Stark shift can be interpreted as a field induced shift of the emitter level splitting. For spin levels this is the equivalent of an effective magnetic field B_{Stark} that creates a Zeeman splitting. Taking into account the two circularly polarized modes ($+$ and $-$) in a Schrieffer–Wolff transformation as shown above, it is found that the effective Zeeman splitting depends on the difference in the number of right- and left-circularly polarized photons,

$$g \mu_B B_{\text{Stark}} = \frac{\hbar \Omega_i^2}{\delta_i} \left(a_+^\dagger a_+ - a_-^\dagger a_- \right). \tag{5.57}$$

This equation provides the intriguing possibility to apply effectively a magnetic field pulse just by a pulse of circularly polarized light. In Section 7.3.8 we elaborate on this topic and review a recent experiment in which such tipping pulses were applied to a single electron spin confined to a quantum dot.

In the following section we explore how cavity photons can couple two emitters or spins.

5.6.2
Two Emitters Interacting with a Cavity

It is straightforward to extend the above considerations to two emitters (or qubits) $i = 1, 2$ that are coupled to the cavity mode. For simplicity, we assume $\Omega_1 = \Omega_2 = \Omega \neq 0$ and that the emitters do not interact by any other mechanism. The Hamiltonian is now given by

$$H = \hbar \sum_{i=1,2} \left[\Delta_i s_{i,z} + \Omega_i \left(a s_i^+ + a^\dagger s_i^- \right) \right] + \hbar \omega a^\dagger a , \tag{5.58}$$

and can be integrated out to leading order by a similar Schrieffer–Wolff transformation $U = \exp A$ as for a single emitter, where now $A = A_1 + A_2$. The resulting effective Hamiltonian is [181, 197]

$$\begin{aligned}
H' = \hbar &\left(\omega + \frac{2\Omega^2}{\delta_1} s_{1,z} + \frac{2\Omega^2}{\delta_2} s_{2,z} \right) a^\dagger a \\
&+ \hbar \left(\Delta_1 + \frac{\Omega^2}{\delta_1} \right) s_{1,z} + \hbar \left(\Delta_2 + \frac{\Omega^2}{\delta_2} \right) s_{2,z} \\
&+ \frac{\hbar}{2} \left(\frac{\Omega^2}{\delta_1} + \frac{\Omega^2}{\delta_2} \right) \left(s_1^+ s_2^- + s_1^- s_2^+ \right),
\end{aligned} \tag{5.59}$$

with $\delta_{1,2} = \Delta_{1,2} - \omega$. The last of the above terms shows that the cavity mediates an effective coupling between the two emitters in second order as it captures processes where a photon is emitted by one emitter and absorbed by the other. Based on this type of interaction, two-qubit gates have been proposed for spins in quantum dots that interact with an optical cavity mode via stimulated Raman transitions [181] and for Cooper pair boxes in a cavity [197]. It is possible to go a step further from here and include an additional direct coupling between the two qubits. This serves as a starting point to investigate how far such an interaction could actually disturb the cavity-mediated coupling, or under which conditions the ac Stark effect can also be used to read out coupled two-qubit states [205].

6
Spin–spin Interaction in Quantum Dots

In this chapter we discuss various types of spin interaction in quantum dots. This involves spins from electrons and holes, as well as nuclear spins. The interaction of electrons includes the dipolar interaction of their spins as well as the Coulomb exchange interaction, which is an effective spin–spin interaction. The electron–hole exchange interaction in an anisotropic environment mixes and reorders the exciton states, which has a practical impact on the transfer of spin quantum states onto photon polarization states and vice versa. Finally, the hyperfine interaction couples the system of electron spins to the nuclear spins embedded in the crystal lattice, which form an important reservoir for the damping of the electron spin dynamics in quantum dots.

6.1
Electron–Electron–Spin Interaction

The interaction of electrons includes the Coulomb interaction as well as the dipolar interaction of their magnetic moments. Regarding the spin dependent ordering of the energy eigenstates, it turns out that both of those interactions contribute, but quite on a different scale.

The Coulomb interaction couples the orbital part of a pair of two electrons. According to the rules of angular momentum addition, the two electron spins S_1 and S_2 can either form an antisymmetric state with total spin 0, the spin singlet state,

$$|S\rangle = \frac{1}{\sqrt{2}} (|\uparrow\downarrow\rangle - |\downarrow\uparrow\rangle), \tag{6.1}$$

or one of the three symmetric states with total spin 1, the spin triplet states,

$$|T_0\rangle = \frac{1}{\sqrt{2}} (|\uparrow\downarrow\rangle + |\downarrow\uparrow\rangle), \quad |T_+\rangle = |\uparrow\uparrow\rangle, \quad |T_-\rangle = |\downarrow\downarrow\rangle, \tag{6.2}$$

for which the z projections are 0, +1, and −1, respectively.

The postulate of Fermi–Dirac statistics that the total wave function must be completely antisymmetric, then implies that a spin singlet has a symmetric orbital

Spins in Optically Active Quantum Dots. Concepts and Methods.
Oliver Gywat, Hubert J. Krenner, and Jesse Berezovsky
Copyright © 2010 WILEY-VCH Verlag GmbH & Co. KGaA, Weinheim
ISBN: 978-3-527-40806-1

wavefunction, while a spin triplet has an antisymmetric orbital wave function. This different orbital symmetry can lead to different Coulomb exchange energies. The resulting energy splitting of singlet and triplet states is in an indirect relation with spin due to symmetry; the Coulomb exchange interaction may thus be considered as an *effective* spin interaction of two electrons. For two spin 1/2 electrons, this effective spin interaction can be represented in the form of a Heisenberg interaction,

$$H_{\text{Heisenberg}} = J\mathbf{S}_1 \cdot \mathbf{S}_2 , \tag{6.3}$$

with the exchange splitting J between the singlet and triplet states, which is easily seen after a transformation into the singlet–triplet basis. In a system of two coupled quantum dots, which we study in more detail in Chapter 8, J is on the order of the interdot Coulomb interaction. Depending on the particular system, J may reach a magnitude of ~0.1 meV [11, 205].

For the magnetic dipolar interaction we consider the magnetic moments $g_1\mu_B\mathbf{S}_1$ and $g_2\mu_B\mathbf{S}_2$, which are associated with the two spins, where μ_B is the Bohr magneton and g_1 and g_2 are the gyromagnetic factors of the two electrons, respectively. The dipolar interaction is given by the expression

$$H_{\text{dipolar}} = \frac{g_1 g_2 \mu_B^2}{R^3} \left[\mathbf{S}_1 \cdot \mathbf{S}_2 - 3\frac{(\mathbf{R} \cdot \mathbf{S}_1)(\mathbf{R} \cdot \mathbf{S}_2)}{R^2} \right], \tag{6.4}$$

where $\mathbf{R}$ is the vector connecting the sites of the two spins. We notice that the dipolar interaction consists of two terms, an isotropic part of the Heisenberg form and an anisotropic part. If we consider two spins that are fixed at two sites and choose our coordinate system such that $\mathbf{R} = R(\sin\theta, 0, \cos\theta)$, then we obtain for the dipolar interaction the more intuitive form

$$H_{\text{dipolar}} = \frac{g_1 g_2 \mu_B^2}{R^3} [\mathbf{S}_1 \cdot \mathbf{S}_2$$
$$-3\left(\sin\theta\, S_{1,x} + \cos\theta\, S_{1,z}\right)\left(\sin\theta\, S_{2,x} + \cos\theta\, S_{2,z}\right)]. \tag{6.5}$$

If we consider spins with a quantization axis along z, then it can easily be seen in the anisotropic part, the second line in above equation, that spin mixing is induced by nonzero terms proportional to the x components of the spin operators. Obviously, the relative spin alignment due to the anisotropic part of the dipolar interaction depends on the relative spatial configuration, θ, of the two spins. The magnetic dipolar interaction of two electron spins in neighboring quantum dots is typically on the order of $\sim 10^{-9}$ meV and therefore much smaller than the exchange interaction.

6.2
Electron–Hole Exchange Interaction

A Coulomb exchange term also appears in the Coulomb interaction matrix elements of electron and hole states in different bands. The inter-band Coulomb exchange interaction, which is usually referred to as the electron–hole exchange interaction, leads to a fine structure splitting of excitonic levels.

6.2.1
Exciton Fine Structure

We review the electron–hole exchange interaction here, considering hh and lh exciton states. According to the method of invariants, the exchange splitting of the ground state of a confined exciton with symmetry $\Gamma_6 \times \Gamma_8$ (see Section 3.1) can be written as an effective coupling of the electron spin $\mathbf{S}$ and the hole angular momentum $\mathbf{J}$ [124, 206],

$$H_{\text{exc}} = -\sum_{i=x,y,z} \left(a_i J_i^h S_i^e + b_i \left(J_i^h \right)^3 S_i^e \right). \tag{6.6}$$

Here, a_i and b_i are coupling constants, and the $j = 3/2$ angular momentum operators J_i for the valence band holes are given by Eqs. (3.18–3.20) as shown in Chapter 3. In a quantum dot with anisotropic shape we expect anisotropic coupling constants a_i and b_i. In the following, we show how such an anisotropic electron–hole exchange interaction affects the exciton states in the basis of the eight energetically lowest bright and dark hh and lh exciton states, denoted by $|J_z^h, S_z^e\rangle$. For our basis we use the exciton states $|\frac{3}{2}, -\frac{1}{2}\rangle, |-\frac{3}{2}, \frac{1}{2}\rangle, |\frac{3}{2}, \frac{1}{2}\rangle, |-\frac{3}{2}, -\frac{1}{2}\rangle, |\frac{1}{2}, \frac{1}{2}\rangle, |-\frac{1}{2}, -\frac{1}{2}\rangle, |\frac{1}{2}, -\frac{1}{2}\rangle, |-\frac{1}{2}, \frac{1}{2}\rangle$, in which Eq. (6.6) takes on the matrix form

$$H_{\text{exc}} \doteq \begin{pmatrix} A_{hh} & A_{hh-lh} \\ A_{hh-lh}^* & A_{lh} \end{pmatrix}, \tag{6.7}$$

where A_{hh}, A_{lh}, and A_{hh-lh} are 4×4 block matrices. Here we note that these combinations of electron and hole states include both optically active (bright) and optically inactive (dark) exciton states. Explicitly,

$$A_{hh} = \begin{pmatrix} \Delta_0 & \Delta_1 & 0 & 0 \\ \Delta_1 & \Delta_0 & 0 & 0 \\ 0 & 0 & -\Delta_0 & \Delta_2 \\ 0 & 0 & \Delta_2 & -\Delta_0 \end{pmatrix}, \tag{6.8}$$

where

$$\Delta_0 = \frac{3}{4} a_z + \frac{27}{16} b_z \tag{6.9}$$

$$\Delta_1 = -\frac{3}{8}(b_x - b_y) \tag{6.10}$$

$$\Delta_2 = -\frac{3}{8}(b_x + b_y). \tag{6.11}$$

The term Δ_1 couples the bright and Δ_2 the dark hh exciton states, respectively. The bright and the dark hh excitons are split by $2\Delta_0$. The block matrix

$$A_{hh-lh} = \begin{pmatrix} \Delta_3 & 0 & 0 & 0 \\ 0 & \Delta_3 & 0 & 0 \\ 0 & 0 & \Delta_4 & 0 \\ 0 & 0 & 0 & \Delta_4 \end{pmatrix} \tag{6.12}$$

mixes hh and lh excitons with

$$\Delta_3 = -\frac{\sqrt{3}}{4}(a_x + a_y) - \frac{7\sqrt{3}}{16}(b_x + b_y), \tag{6.13}$$

$$\Delta_4 = -\frac{\sqrt{3}}{4}(a_x - a_y) - \frac{7\sqrt{3}}{16}(b_x - b_y), \tag{6.14}$$

and, finally, the coupling of the lh exciton states is provided by

$$A_{lh} = \begin{pmatrix} -\Delta_5 & \Delta_6 & 0 & 0 \\ \Delta_6 & -\Delta_5 & 0 & 0 \\ 0 & 0 & \Delta_5 & \Delta_7 \\ 0 & 0 & \Delta_7 & \Delta_5 \end{pmatrix}, \tag{6.15}$$

where the splitting between dark and bright lh excitons amounts to $2\Delta_5$, and

$$\Delta_5 = \frac{1}{4}a_z + \frac{1}{16}b_z, \tag{6.16}$$

$$\Delta_6 = -\frac{1}{2}(a_x - a_y) - \frac{5}{4}(b_x - b_y), \tag{6.17}$$

$$\Delta_7 = -\frac{1}{2}(a_x + a_y) - \frac{5}{4}(b_x + b_y). \tag{6.18}$$

Let us first have a look at the ideal case. For rotational symmetry of the dot, and for cubic crystal symmetry (which is given for a $\Gamma_6 \times \Gamma_8$ exciton) we obtain $a_x = a_y$ and $b_x = b_y$. In this case, H_{exc} is diagonal for the bright excitons ($\Delta_1 = \Delta_6 = 0$), and we obtain bright excitons according to the usual interband transitions. In contrast, the dark excitons are mixed due to Δ_2 and Δ_7. In quantum dots with a splitting Δ_{hh-lh} of heavy and light hole states, the electron–hole exchange interaction can further be simplified [206, 207]. As for quantum dots typically $\Delta_{hh-lh} \sim 10$ meV, the hh–lh splitting is much larger than typical electron–hole exchange interaction energies ($\lesssim 0.05$ meV in "natural" GaAs dots [24], $\lesssim 0.2$ meV in InAs dots [168, 207, 208], $\lesssim 0.3$ meV in CdSe/ZnSe dots [209]). We may therefore restrict ourselves to hh states and the exchange interaction described by Eq. (6.8).

For a quantum dot with an asymmetric shape in the xy plane, $b_x \neq b_y$ is possible, leading to a coupling of $|\frac{3}{2}, -\frac{1}{2}\rangle$ and $|-\frac{3}{2}, \frac{1}{2}\rangle$. The circularly polarized bright excitons thus transform into linearly polarized exciton states,

$$\frac{1}{\sqrt{2}}\left(\left|\frac{3}{2}, -\frac{1}{2}\right\rangle \pm \left|-\frac{3}{2}, \frac{1}{2}\right\rangle\right) \tag{6.19}$$

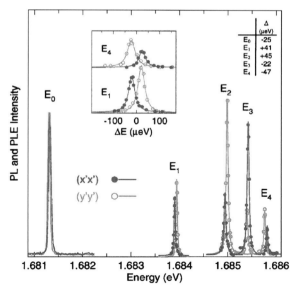

Fig. 6.1 PL excitation spectrum of a single neutral IFQD with no applied magnetic field. The $x'x'$ ($y'y'$) spectrum is measured with both the excitation light and the collected light polarized linearly along the [110] ($[\bar{1}10]$) directions. Reprinted with permission from [24]. Copyright (1996) by the American Physical Society.

which are split by the energy $\delta_{\mathrm{ehx}} = 2\Delta_1$. For elliptic dots these intrinsic basis states are oriented along the major and the minor axis, respectively [210].

For example, GaAs interface fluctuation quantum dots tend to be elongated along the $[\bar{1}10]$ crystal direction (see Figure 2.3 in Chapter 2). In this type of dots the electron–hole exchange interaction results in a fine-structure splitting δ_{ehx} of the exciton PL peak on the order of tens of μeV, as shown in Figure 6.1. If a magnetic field is applied such that the Zeeman splitting is larger than this anisotropic exchange splitting, then the circularly polarized eigenstates are restored [211].

The mixing of bright exciton spin states in anisotropic quantum dots is undesirable for a faithful conversion of quantum information from spin states into photon polarization states, and vice versa. In this context, an intriguing idea is the generation of a pair of polarization-entangled photons from the decay of a ground state biexciton $2X^0$ in a quantum dot. In the following section we introduce this concept and ways how the exciton fine structure can be overcome in practical applications.

6.2.2
Biexcitons and Polarization-Entangled Photons

The decay of a ground state biexciton in a quantum dot has been proposed to generate polarization-entangled photons [149, 212]. This scheme is in some sense analogous to atomic *s-p-s* cascades, for which photon entanglement has been demonstrated by several violations of Bell's inequalities [213–215].

The exciton fine structure plays a crucial role in the realization of this biexciton proposal. The decay of a biexciton in a quantum dot occurs typically in a cascade of two subsequently emitted photons, each from a recombining electron–hole pair. It turned out in a number of experiments that the photons were not entangled if the exciton, which remained after the emission of the first photon in the quantum dot, was subject to a resolvable fine structure splitting [216–220]. This exciton energy splitting effectively erased the quantum mechanical interference of the two decay paths that was necessary to transfer the exciton spin entanglement into a photon polarization entanglement. The splitting of the intermediate exciton spin states meant that the photon that was emitted first and the remaining exciton in the dot were not only entangled in spin and polarization, respectively, but also in energy or frequency. Hence, the entanglement spread out into the orbital part of the exciton where any measurement of the energy would already project the entangled wavefunction into an unentangled state.

Two groups then reported the successful generation of polarization-entangled photons from the biexciton cascade in a single quantum dot [221, 222]. Improving the quantum dot growth conditions in both cases led to a reduced exciton splitting, and spectral filtering then erased the "which path" information of the decay cascade.

Some alternative proposals involving spin and photon entanglement have been made to overcome the problem with the exciton fine structure splitting in quantum dots. These include applying an in-plane magnetic field to merge the exciton lines by a Zeeman shift, and also making use of charged excitons. For example, a scheme using two positively charged excitons X^{1+} in two quantum dots for the generation of two polarization-entangled photons has been proposed, which can also be extended to a scheme producing entangled four-photon states of the Greenberger–Horne–Zeilinger type [223]. A recent proposal to electrically generate entangled electron spin-photon states using interband transitions of a double dot in a *p-i-n* diode structure is based on negatively charged excitons, X^{1-}, which have a vanishing fine structure splitting as the electrons couple to a spin singlet [224].

We have shown in Section 5.2.2 that the photon polarization depends on the emission direction with respect to the exciton angular momentum quantization axis z. This also affects the degree of entanglement of the two photons. Gywat *et al.* have calculated the von Neumann entropy to quantify the degree of bipartite photon entanglement of the biexciton cascade as a function of the photon emission angles [139, 223]. As a result, the entanglement is maximal for emission along z. For emission at small polar angles θ around z, entanglement is only weakly reduced with θ, but then drops to zero if any of the two photons is emitted perpendicular to z. Surprisingly, due to quantum mechanical interference, a maximum photon entanglement can be obtained by choosing appropriate azimuthal detection angles, for which the admixed components with opposite circular polarization cancel. This leaves behind the original maximally entangled state for all angles θ except for $\theta = \pi/2$, for which destructive interference is maximal [223].

6.3
Hyperfine Interaction

The Hamiltonian for the magnetic interaction between an electron with spin $\mathbf{S}$, gyromagnetic factor g ($g = 2.00$ for a free electron), and orbital angular momentum $\mathbf{L}$, and a nucleus with spin $\mathbf{I}$ and gyromagnetic factor γ is given by (see [225])

$$H_{\mathrm{HF}} = g\mu_{\mathrm{B}}\gamma\hbar\mathbf{I} \cdot \left[\frac{\mathbf{L}}{r^3} - \frac{\mathbf{S}}{r^3} + 3\frac{\mathbf{r}(\mathbf{S}\cdot\mathbf{r})}{r^5} + \frac{8\pi}{3}\mathbf{S}\delta(\mathbf{r}) \right]. \tag{6.20}$$

The second and third terms form the usual magnetic dipole–dipole interaction term, as has already been shown for the electron–electron dipolar interaction. The term with the delta function $\delta(\mathbf{r})$ is called the Fermi contact hyperfine interaction. For an atomic electron with a total angular momentum $j = l \pm 1/2$ it is possible to write $H_{\mathrm{HF}} = A_j\mathbf{I}\cdot\mathbf{j}$. Here, for an s electron we obtain

$$A_j = \frac{8\pi}{3}g\mu_{\mathrm{B}}\gamma\hbar\,|\psi(0)|^2 \,, \tag{6.21}$$

while for $l \neq 0$,

$$A_j = g\mu_{\mathrm{B}}\gamma\hbar\left\langle\frac{1}{r^3}\right\rangle\frac{l(l+1)}{j(j+1)} \,, \tag{6.22}$$

with the expectation value $\langle\ldots\rangle$. Similarly, for a free atom or a paramagnetic molecule with many electrons with a total angular momentum J it is possible to write $H_{\mathrm{HF}} = A_j\mathbf{I}\cdot\mathbf{J}$.

The hyperfine interaction leads to a number of interesting phenomena since the two classes of spins involved have quite different properties. Nuclear spins couple much more weakly to their environment than electron spins, as their main interaction types are the Fermi contact hyperfine interaction with electrons, as well as the dipolar interaction with spins of spatially close electrons and other nuclei. Nuclear spins typically have much longer lifetimes than electron spins, which are much more strongly coupled to their environment.

The hyperfine interaction energy shows up as characteristic shifts of the electronic and nuclear energy levels if there is polarization of one part of the spin system. For the electrons, an energy shift induced by nuclear spin polarization is called the Overhauser shift. In turn, a shift of nuclear energy levels due to electron spin polarization is referred to as the Knight shift.

The presence of hyperfine levels that involve different configurations of electron and nuclear spins also provides powerful spin relaxation and decoherence processes in some systems due to mutual spin flips of electrons and nuclei. These effects are usually strong at zero external magnetic field where spin flips are energy conserving. Yet, due to the different magnetic moments of nuclei and electrons, a mismatch of their Zeeman splittings can be induced by a finite magnetic field, which quenches this interaction channel.

For an electron in a typical semiconductor quantum dot, the electron wave function may easily extend over $\sim 10^5$ nuclei. In gallium arsenide, for instance, each of these nuclei has spin $I = 3/2$. The contact hyperfine interaction then provides the dominant coupling of electron and nuclear spins. As the electron probability density changes across the quantum dot, the coupling strength to the nuclei varies from typically about $A_j \approx 10^{-6}$ meV near the dot center down to $A_j \approx 0$ at the tails of the electron wave function [11]. As a result of the hyperfine interaction, spin coherence of electrons in most types of quantum dots is suppressed efficiently and typically lost on the order of ~1 ns. Several publications have addressed the theory of hyperfine-induced electron spin decoherence [138, 226–228]. A more detailed overview of this topic can be found in [11].

Recently, hole spins in quantum dots have received increasing attention as carriers of quantum information. See Section 7.2.2 for a few recent experiments in this direction. Basically, for the v band the hyperfine interaction is different than for the c band. First of all, the Fermi contact hyperfine interaction vanishes because the p-like Bloch functions have zero overlap with the nuclear sites. Further, the dipolar coupling term for heavy holes takes in the quasi two-dimensional limit the form of a simple z–z or Ising coupling, as shown by Fischer *et al.* [229]. These authors have shown that in spite of these apparent simplifications, the hyperfine interaction of the hole spin can be quite strong, and the hyperfine coupling strength can be about only an order of magnitude smaller than for the c band. This analysis holds for unstrained material. The Ising form of the hyperfine coupling in this regime means that the main source of decoherence for hole spins is the broad frequency distribution of the nuclear spins (due to the absence of the hyperfine flip-flop terms) [229].

7
Experimental Methods for Optical Initialization, Readout, and Manipulation of Spins in Quantum Dots

Much has been written about the potential for quantum information processing in various systems. In virtually all cases, the requirements that must be met for any useful operations to be carried out can be divided into three categories. First the quantum system that is being used as a qubit must be *initialized* into a known state. In the case of electron spin qubits, for example, all of the spins might be placed into a spin "up" state. Next the actual quantum operations must be carried out. In general, this may involve operations on a single qubit, or operations that couple two or more qubits together. In both cases, the qubit states are *manipulated*. For spin qubits, a single qubit operation may be realized by coherent control of the spin state, or a two-qubit operation may be effected by letting two spins evolve under their mutual exchange interaction. Finally, usually after these manipulations have been completed, the final states of the qubits must be *read out*. This corresponds to a measurement of spin "up" or spin "down" for each qubit in the case of electron spin qubits.

These three requirements – initialization, manipulation and readout – are all possible using optical methods for a qubit consisting of an electron spin in a quantum dot. To be sure, many challenges remain before a useful quantum device can be made based on optically controlled spins in quantum dots. But individually, the ingredients necessary have all been demonstrated in a variety of ways.

Optical initialization of spins in quantum dots can be quite straightforward, simply relying on the spin dependent optical selection rules often present in semiconductors. More involved spin pumping schemes, as discussed below, can be used to achieve higher initialization fidelity. The same selection rules that allow for spin initialization also permit the spin state to be read out – most simply, by the polarization of light emitted upon recombination of an electron and a hole. Additionally, spin dependent absorption or the Faraday effect may be used to read out the spin. Finally, nonlinear optical effects such as Raman transitions or the optical Stark effect may be used to optically manipulate the spin of electrons in a quantum dot.

This chapter will describe various experimental methods for initialization, manipulation, and readout of spins in quantum dots by highlighting recent results from the literature. This is not intended to be an exhaustive review, but instead to serve as an illustration of the various techniques being employed.

Spins in Optically Active Quantum Dots. Concepts and Methods.
Oliver Gywat, Hubert J. Krenner, and Jesse Berezovsky
Copyright © 2010 WILEY-VCH Verlag GmbH & Co. KGaA, Weinheim
ISBN: 978-3-527-40806-1

Table 7.1 Main results of papers highlighted in this chapter, with key figures of merit in bold.

Optical Orientation [121]	Spin pumping in bulk semiconductors
Cortez *et al.* [230]	Nonresonant spin pumping in dots
Atatüre *et al.* [161]	Resonant spin pumping in single dots **High fidelity: >99.8%**
Gammon *et al.* [231]	Dynamic nuclear polarization in quantum dots
Kroutvar *et al.* [232]	Spin storage and readout $T_1 \approx 20$ **ms**
Besombes *et al.* [233]	Readout of a single Mn spin in a quantum dot
Högele *et al.* [234]	Single dot spin-selective absorption
Epstein *et al.* [235]	Hanle measurement on an ensemble of dots
Bracker *et al.* [211]	Single dot Hanle measurement $T_2^* \approx 15$ **ns**
Gupta *et al.* [236]	Time-resolved Faraday rotation on nanocrystal dots $T_2^* \approx 2$ **ns at 300 K**
Greilich *et al.* [237]	Spin echoes in a quantum dot ensemble $T_2 \approx 3$ **µs**
Berezovsky *et al.* [60]	Single spin Kerr rotation measurement
Mikkelsen *et al.* [238]	Observation of coherent dynamics of a single spin
Berezovsky *et al.* [239]	Ultrafast manipulation of spin coherence
Press *et al.* [240]	Coherent spin control, Ramsey fringes **Up to 13 π rotations in several ps, 91% fidelity π-rotation**

7.1
Optical Spin Initialization

The groundwork for the optical initialization of spins in quantum dots goes back to the work on optical orientation of spins in bulk semiconductors, mostly done in the 1960s and 1970s. These results are presented in the seminal book, *Optical Orientation* [121], published in 1984. There, the selection rules for interband transitions in semiconductors are exploited to selectively excite spin-polarized electrons and holes into the conduction and valence bands, respectively (see also Chapter 5). To review this effect, a schematic of the band structure of GaAs is shown in Figure 7.1a. Optical transitions are possible from the heavy hole band (angular momentum $J = 3/2$, $J_z = \pm 3/2$), the light hole band ($J = 3/2$, $J_z = \pm 1/2$), and the split-off band ($J = 1/2$, $J_z = \pm 1/2$). For the purposes of optical orientation, the energy of light is typically chosen to not excite carriers from the split-off band. Figure 7.1b shows the transitions involving circularly polarized light from the heavy and light hole bands to the conduction band. Since the absorption of a circularly polarized photon must be accompanied by a change of angular momentum $\pm \hbar$ elsewhere, an electron–hole pair created by such absorption can only take the paths as indicated in the figure. Since the probabilities for the heavy hole transition is three times larger than that of the light hole transition, circularly polarized light that excites both transitions yields an ensemble of electrons with an average spin

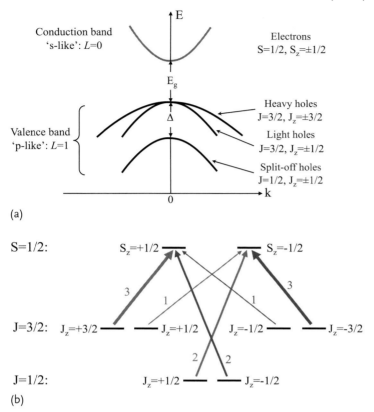

Fig. 7.1 (a) Band structure of GaAs; E_g is the band gap, Δ the spin–orbit energy. (b) Circularly polarized band edge optical transitions, with relative strength of each transition.

polarization of ideally 50%. If strain is introduced into the semiconductor lattice, then the heavy–light hole degeneracy is lifted. Then the energy of the optical excitation may be tuned to only excite heavy or light holes. In this case, up to 100% spin polarization may be obtained.

If spin polarized electrons and holes are excited in an undoped semiconductor they will recombine on a timescale typically less than a nanosecond, with the polarization of the emitted light related to the spin polarization of the pair at the time of recombination. The results of *Optical Orientation* also discuss optical spin polarization in doped semiconductors. In this case, the spins of the dopant carriers (for example, consider electrons) can also be optically polarized. When circularly polarized light is used to generate excess spin-polarized electrons and holes, the holes may recombine with unpolarized dopant electrons. In fact, this will predominantly be the case if the dopant electrons outnumber the optically excited ones. When this occurs, then the spin-polarized electron is left behind in the conduction band.

This allows observation of the electron spin over periods of time not limited by the recombination of electrons and holes [241].

These results for optical spin initialization in bulk semiconductors largely hold true for spins confined to quantum dots, with a few complications here and there. When a spin polarized electron–hole pair is excited in a quantum dot system, the charge carriers may be created within the dot itself, in a "wetting layer" or even in the surrounding barriers. (Here, the wetting layer refers to the wetting layer present in Stranski–Krastanov dots, or the quantum well in interface fluctuation dots – see Chapter 2.) Once excited, the electron and hole can then relax down into the lowest available state of the dot. When the dot is unoccupied, this energy relaxation goes all the way down to the ground state, typically maintaining the spin state to some extent (at least for the electron, the hole spin is often lost). Thus one can pump spin polarized excitons fairly easily. One exception to this rule is in dots with significant shape anisotropy, which causes electrons and holes to relax into linearly polarized eigenstates [24].

Of particular interest are processes that can polarize a single electron spin in a quantum dot, with the absence of a hole. The electron–hole recombination time is typically on the order of 100 ps in many dots. The electron spin lifetime, on the other had, can be significantly longer. Therefore by looking at single electrons, the spin can be studied over much longer timescales.

7.1.1
Nonresonant Spin Pumping

In order to observe the long-term dynamics of single electron spins in quantum dots, one must first dope the quantum dots such that they contain a single electron in equilibrium. This can be accomplished by incorporation of the dots into a charge-tunable electronic structure (see Chapter 4), or by including an n-type delta doped layer near the quantum dots. It is this second approach that is used in the work of S. Cortez *et al.*, "Optically driven spin memory in n-doped InAs-GaAs quantum dots" [230]. In these experiments, circularly polarized nonresonant excitation is used to polarize the spin of the dopant electron in self-assembled InAs quantum dots.

The self-assembled quantum dots in this study are grown by molecular beam epitaxy (see Chapter 2), and the doping concentration is chosen to add approximately one electron to each dot. The energy of the excitation laser is tuned to the InGaAs wetting layer (1.44 eV), whereas the photoluminescence of the dots is centered around 1.15 eV. By changing the helicity of the excitation (either σ^+ or σ^-), either spin up or spin down electrons may be added to the continuum of states in the wetting layer. The optically excited electrons and holes then rapidly relax into the quantum dots. The singly doped dots then contain the two electrons and a hole, in a "trion" or negatively-charged exciton state. One of the electrons may then recombine with the hole, emitting a photon whose circular polarization depends on the spin of the recombining carriers. Figure 7.2a illustrates the initial trion and final single electron state of the dot. Assuming the electrons are initially in the sin-

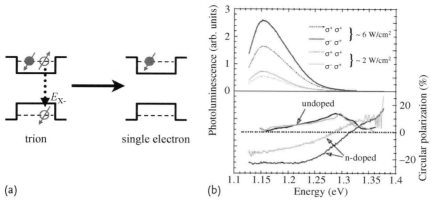

Fig. 7.2 (a) Decay of a trion leaves a single electron in the dot. (b) *Top:* Polarized luminescence spectra of a quantum dot ensemble at two pump intensities. *Bottom:* Polarization of luminescence spectra in undoped and n-doped dot ensembles. Reprinted with permission from [230]. Copyright (2002) by the American Physical Society.

glet ground state, the spin of the electron remaining in the dot is then opposite to that of the annihilated electron. If the dynamics of the situation are such that electrons of one spin orientation are more likely to recombine with the hole, then there will be a net spin polarization of the single electrons left behind.

The upper panel of Figure 7.2b shows the photoluminescence spectrum of the ensemble of n-doped InAs dots pumped with both σ^+-polarized excitation in the wetting layer, and with the σ^+- and σ^--polarized PL collected separately. The polarization of the PL can be defined as

$$P = \frac{I_{++} - I_{+-}}{I_{++} + I_{+-}} , \tag{7.1}$$

where I_{++} (I_{+-}) is the intensity of the σ^+ (σ^-) luminescence pumped with σ^+-polarized excitation. The lower panel of Figure 7.2b shows the polarization of the PL in the top panel, as well as the polarization of PL from a sample of undoped dots. (The experimental setup for these measurements, as well as additional time-resolved measurements from this work will be discussed further below.) The positive polarization of the undoped dots simply reflects the polarization of the optically oriented electrons and holes. The situation is more complicated for the dots doped with an additional electron. The PL from the singly-charged dots displays polarization opposite to that of the excitation light. To understand this result, one must take into account the various pathways for the relaxation of carriers in the dot.

It is assumed in this work that the polarization of the holes is lost before being captured by the dots due to the strong spin–orbit coupling in the valence band, and the initial polarization of the electrons is essentially 100%. When an electron and hole relax into a dot, there are two possibilities: the spin of the resident electron in the dot is parallel or anti-parallel to the spin of the optically excited electron. If they are anti-parallel, then the optically-excited electron can relax all the way down to the singlet ground state of the trion. The subsequent recombination of the hole with one of the two electrons then depends on the randomly oriented hole polarization, and the luminescence is on average unpolarized. In the case where the electrons are parallel however, as long as the spin of the injected electron is preserved, it can only relax into a higher-energy triplet state. The possible relaxation pathways are illustrated in Figure 7.3.

If the electron relaxes into this triplet state, there are two further possibilities: that the hole spin is polarized parallel or anti-parallel to the two electron spins. In the latter case, the resident electron in the lowest energy level may just directly recombine with the hole, resulting in positive polarization. However, there is another strong relaxation pathway in these dots due to the anisotropic exchange interaction between the electrons and holes. This interaction allows the higher energy electron to rapidly relax to the singlet ground state by way of a spin flip with the hole. The emission from this state is then polarized opposite to the excitation polarization. When the hole spin is initially polarized parallel to the electron spins in the trion state, no recombination can occur, and the anisotropic exchange mechanism is not significant in this state. Eventually, the higher energy electron can relax into the ground state via a phonon-assisted transition, also yielding negatively polarized emission. By analyzing the rates of these various mechanisms, Cortez *et al.* found that the negatively-polarized emission dominates, as is observed experimentally.

The conclusion of the arguments above is that the nonresonant excitation of spin polarized electrons and holes into singly-charged quantum dots pumps the

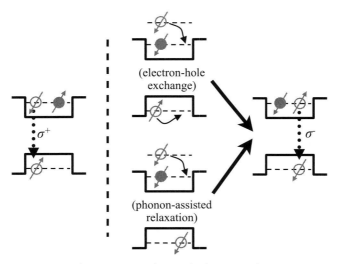

Fig. 7.3 Trion decay processes that can lead to positively or negatively circularly polarized emission. When the resident spins are preferentially polarized in the down direction, the processes on the right dominate.

spin of the resident electron in the dot in a direction parallel to that of the injected electrons. However, because this polarization results from a competition between positive and negative contributions, the magnitude of the polarization is limited, in this case, to about −20%. Similar effects are observed in GaAs interface fluctuation quantum dots, though are equally complicated to explain.

7.1.2
Resonant Spin Pumping

The limit of the maximum achievable spin pumping by nonresonant excitation can be overcome by using resonant spin pumping schemes. For example, in the work of M. Atatüre *et al.*, "Quantum-dot spin-state preparation with near-unity fidelity" [161], the spin polarization of an electron in a quantum dot is polarized to better than 99.8%.

In this work, the sample consists of self-assembled InAs quantum dots embedded in a diode structure, which allows for electrical control of the equilibrium charge state of the dots. The quantum dot here is operated in the singly charged state. Additionally, the ability to apply an electric field across the structure gives one fine control over the energy levels via the quantum confined Stark shift. This experiment is performed on a single dot, which is first characterized by microphotoluminescence spectroscopy (see Chapter 2). To measure the spin in the quantum dot, a differential absorption technique is employed. This method measures the optical absorption due to the quantum dot transitions, which can reveal the spin state due to Pauli blocking (see Section 7.2.4 for more details).

A narrow-linewidth continuous-wave laser is used to both probe the absorption of transitions and also to perform the spin pumping. In contrast to the previous example, here the excitation of the dot is performed resonantly. In order to understand the physics of this system, it is useful to look at the different optical transitions in the quantum dot. Figure 7.4a and 7.4b shows these transitions, with the bottom two levels indicating the spin-up and spin-down single electron ground states, and the upper levels indicate the trion states with the electrons in the singlet ground state and either hole spin up or down.

Due to the selection rules previously discussed, circularly-polarized excitation of one helicity strongly drives the transition between the two states on the left, and the other helicity drives the transition between the states on the right. Transitions between the left trion state and the right single electron state (and vice versa) are forbidden to first order, but still occur weakly due to heavy-light hole mixing, or an external magnetic field. Additionally, transitions between the spin-up and spin-down single electron states can occur due to spin flips with nuclear spins or interaction with the nearby electron reservoir. However, these transitions are suppressed at nonzero magnetic field, due to the different Zeeman splittings for the electrons and nuclei.

The spin pumping is achieved by driving one of the allowed transitions using circularly polarized excitation, as shown in Figure 7.4. Primarily, the system transitions between the spin-down state and the hole-spin-down trion state (Figure 7.4a).

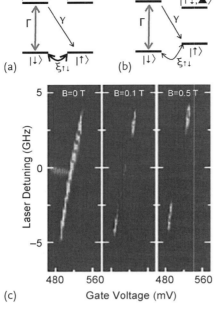

(c)

Fig. 7.4 (a) A circularly polarized laser drives the transition, at a rate Γ, to the trion state. Occasionally with rate γ, this state may relax to the opposite single spin state, which may then flip back via an electron-nuclear spin flip, at a rate $\xi_{\uparrow\downarrow}$. (b) In a magnetic field, $\xi_{\uparrow\downarrow}$ is suppressed and electrons in the spin-up state are trapped. (c) Absorption of the spin-down to trion transition as a function of gate voltage, at various magnetic fields. With a magnetic field present, absorption is suppressed over a range of bias as the electron is trapped in the spin-up state. Reprinted with permission from [163]. Copyright (2008) by the American Physical Society.

Occasionally, the spin-up state may be populated through a "forbidden" transition from the trion, or by a nuclear-induced spin flip. However, the state can return to the spin-down state via another electron-nuclear spin flip, and the bright transition will continue to occur. The situation is different, however, when a magnetic field is applied (Figure 7.4b). In this case, the nuclear-induced spin flips do not occur. Now if the weakly allowed transition takes place from the trion state to the spin up state, there is no way to return to the spin down state – the spin has been pumped into the spin up state. Even though the transition to the spin up state is not too likely, it will happen eventually, and then the spin will have been pumped.

This effect can be observed in the absorption spectra shown in Figure 7.4c. The observed line in the left panel of Figure 7.4c shows the absorption of the transition being driven in part a. (The slope of the line is due to the quantum confined Stark shift of the levels as the gate voltage is swept.) In the middle and right panels of Figure 7.4c, the absorption of this transition is largely suppressed because the state is stuck in the spin-up state and the spin-down trion transition can no longer be driven. The suppression does not occur over the entire singly-charged bias range

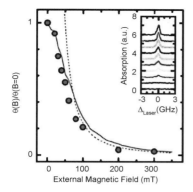

Fig. 7.5 Suppression of the spin-down to trion absorption, θ, as the magnetic field is increased. The measurement sensitivity limits the observed suppression to about 0.002, or a spin preparation fidelity of 99.8%. Inset: Suppression of the absorption peak with spin pumping. Reprinted with permission from [163]. Copyright (2008) by the American Physical Society.

since near the edges, coupling to the electron reservoir in the device leads to spin flips between the two ground states. Similarly, the spin pumping is only efficient for a range of magnetic field, large enough to suppress the nuclear-induced spin flips, and small enough to avoid phonon-related spin flips [163]. The magnitude of this suppression at the optimal conditions is shown in Figure 7.5, as the magnetic field is increased. The absorption is decreased by a factor of about 10^{-3} until the detection sensitivity of this experiment is reached. This means that the quantum dot is in the single spin up state about 99.8% of the time.

A similar method has been used by D. Kim *et al.* [242] to initialize and measure spins in two coupled InAs quantum dots. Here, initialization fidelity of 96% was obtained. Furthermore, due to the coupling of the dots, the spin state of the electron in one of the dots may be read out using a transition involving the other dot. In this way, the spin can be measured nondestructively (that is, without driving transitions involving the initialized spin itself).

One potential drawback of this method of spin pumping is that it can only initialize the spin state into an eigenstate (one of the two Zeeman-split levels). A solution to this problem has been demonstrated by X. Xu *et al.* [243], by using a second laser to trap an electron spin in a quantum dot in an arbitrary coherent state. This "coherent population trapping" operates in a similar way to the scheme describe above, where a laser drives one of the transitions of a three-level system. In this case, however, the transition that is "forbidden" above (that is, slow) is also optically allowed, and is driven by a second laser. The coherent interaction of these two lasers with the three level system drives transitions when the system is in a particular coherent state, but does not drive transitions when the system is in another coherent "dark" state. Thus, the two lasers will drive transitions in the dot until the spin winds up in the dark state, where the spin will then be trapped. The particular state that is trapped can be controlled by adjusting the relative intensity of the two lasers.

Alternatively, from a quantum information processing standpoint, the spin initialization step must then be followed by efficient spin rotations. Thus if the spin is initialized into an eigenstate, one may then simply perform rotations on the spin to produce an arbitrary coherent state. This is the route taken by Press *et al.* [240], as described at the end of this chapter.

Another potential drawback of this pumping technique is that it cannot be performed in zero magnetic field. The field is needed to suppress the nuclear-spin-induced electron spin flips that occur at low magnetic field. This problem has been avoided by B. Gerardot *et al.* [244], by looking at a hole spin in the valence band as opposed to an electron in the conduction band. Because the hole states have *p*-like symmetry, their wavefunctions have a node at the lattice ions. Therefore there is very little coupling between the hole spin and the nuclear spins, and the pumping scheme described above works at zero magnetic field. The lack of interaction with the nuclear spins also makes holes interesting for potentially longer spin dephasing times, as mentioned again in the next section.

7.1.3
Nuclear Spin Pumping

When the electron spins are polarized in a quantum dot, the spin may also be transferred to the surrounding nuclear spins via the hyperfine interaction (see Section 6.3). The polarization of Ga and As nuclear spins of an interface fluctuation dot is studied in the work of D. Gammon *et al.*, "Electron and nuclear spin interactions in the optical spectra of single GaAs quantum dots" [231]. In these experiments, undoped quantum dots are formed at the interfaces of a 4.2-nm-thick AlGaAs/GaAs quantum well. One quantum dot is located and measured using micro-photoluminescence spectroscopy. Electrons and holes are pumped into the quantum well states (analogous to the wetting layer in self-assembled dots) using circularly polarized light. The spin-polarized electrons that relax into the dot interact and undergo spin flips with the nuclear spins. Over several seconds, the electron spins are continuously repumped and build up a substantial polarization of nuclear spins.

Also through the hyperfine interaction, the nuclear spin polarization causes a splitting in the electron levels (the Overhauser shift). This splitting can be directly observed in the photoluminescence spectrum from the quantum dot. Figure 7.6 shows such a spectrum from a single dot. The nonresonant excitation is circularly polarized, and only linearly polarized light along the [110] axis is collected. At zero magnetic field, a single narrow line is seen (the orthogonally polarized luminescence is also shown, with a small splitting seen due to the anisotropic exchange interaction). As a magnetic field is applied, the line splits into a doublet due to the combined effects of the Zeeman splitting and the Overhauser shift.

Figure 7.7 shows the splitting between the two PL lines as a function of applied magnetic field for both right and left circularly polarized excitation. Except for a small region about zero field, the splitting changes linearly with magnetic field as expected for the usual Zeeman splitting. However, the *y*-intercept of the

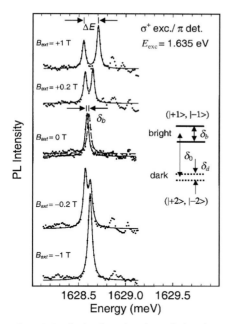

Fig. 7.6 Single dot, linearly polarized photoluminescence spectra, with σ^+ polarized excitation. As a magnetic field is applied, the observed splittings change due to the Zeeman and Overhauser effects. At zero field, both linearly polarized components are shown, with a small splitting due to the anisotropy of the dot shape. Reprinted with permission from [231]. Copyright (2001) by the American Physical Society.

splitting is offset from zero in the positive direction for σ^+ excitation, and in the negative direction for σ^- excitation. This is indicative of nuclear spin polarization, the sign of which depends on the polarization of the optically excited electron spins. The observed splitting corresponds to a polarization of roughly 65% of the nuclei interacting with the electron in the dot. Seen as an effective magnetic field, the nuclear spins are providing a field of $B_N = 1.2\,$T. (Note that the energy shifts of the lines cross zero at $\pm 1.2\,$T when the nuclear polarization cancels the applied field.)

Additional evidence that this offset is due to nuclear spin polarization is obtained by sweeping an applied radio-frequency (RF) field through the nuclear spin resonance frequencies. When the frequency sweeps through the resonance, the nuclear spins are depolarized. The lower inset in Figure 7.7 shows this depolarization as a function of the RF frequency scan rate. From this plot, it can be seen that the nuclei require about 3 seconds to repolarize between depolarization events. This timescale on the order of seconds is typical for dynamic nuclear polarization processes and provides a good indication that the nuclear spins are responsible for the observed effect. Virtually no other spin-related effects in these systems are known to operate on such slow timescales.

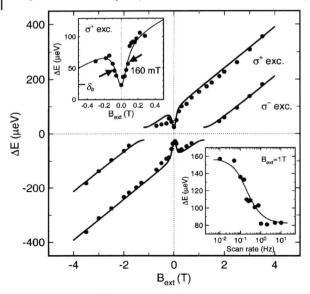

Fig. 7.7 Splitting of the quantum dot luminescence as a function of magnetic field, for excitation of both helicities. The data show the expected linear Zeeman splitting as well as a shift due to the nuclear polarization. Top inset: close-up of the zero-field dip (see [231]).

Bottom inset: Depolarization of the nuclear spins, showing a build-up time of several seconds. Reprinted with permission from [231]. Copyright (2001) by the American Physical Society.

The polarization of nuclear spins is a fairly ubiquitous phenomenon in quantum dots. It has been suggested that nuclear spins can possibly be used as a relatively long term way of storing quantum information. Long time scale ($\sim$ minutes) interactions have been observed between nuclear and electron spin states in quantum dots (see, e.g., Greilich *et al.* [245], discussed in Section 7.3.5). Such "quantum memory" schemes have been demonstrated by transferring quantum information from the electron spin to a single nuclear spin of a phosphorus donor [246], or a defect state (nitrogen-vacancy center) in diamond [247]. The situation is more complicated in quantum dots, however, since an electron in the dot typically interacts with many nuclear spins.

Often, however, nuclear spin polarization is an unwanted complication, and is suppressed by rapidly modulating the helicity of optical excitation. Since the nuclear spins require seconds to polarize, if the excitation helicity is switched much faster than this, the nuclear polarization will remain negligible. Also, as mentioned above, hole spins are expected to interact very weakly with the nuclear spins. Thus, the use of a hole spin as a qubit may by a promising route.

7.2
Optical Spin Readout

In optically active quantum dots, there are several ways to read out the state of spins using optical means. The simplest method is to measure the polarization of luminescence from the quantum dots. Due to spin dependent interband transitions governed by optical selection rules in the presence of the spin-orbit interaction, the polarization of luminescence can often yield information about the spin state in the dot before recombination of the carriers.

One of the simplest schemes for measuring polarized photoluminescence is shown in Figure 7.8. Here, circularly polarized light is focused onto the sample to pump spin polarized carriers as discussed in the previous section. Circular polarization is achieved by passing the excitation light (typically provided by a laser) through a linear polarizer followed by a quarter wave plate with its axis at 45° to the polarizer axis. The quarter wave plate retards the phase of the component of the light along the wave plate axis by a quarter of one wavelength. The result is that, in this configuration, the linearly polarized light is converted to either right or left circular polarization depending on whether the quarter wave plate is set to plus or minus 45° with respect to the linear polarizer. (For a detailed discussion of polarization optics, see [174].) Any excitation light that is transmitted through the sample is then blocked to prevent this light from entering the detector.

Assuming the luminescence is emitted isotropically, the light collection path can be placed at any angle to the excitation. However, the spins will be polarized along the excitation light direction, so if one wishes to measure the polarization along the same axis, the excitation and the collection paths should be as parallel as possible. The luminescence is collected and collimated by a lens whose focal point is over-lapped with the excitation spot. Emitted light of one circular polarization can be selected by repeating the same operation as in the excitation path in reverse. A quarter wave plate converts the circularly polarized components of the light to orthogonal linear polarization components. A subsequent linear polarizer then transmits only one such linear polarization. The transmitted light is then sent to a detector, such as a photodiode. In order to measure the polarization as defined in Eq. (7.1), one must measure the intensity at the detector, then rotate the linear polarizer in the collection path by 90° to measure the other circular component of the emission. Alternatively, the quarter wave plate in the excitation path may be rotated to switch the helicity of the optical excitation.

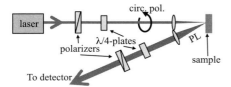

Fig. 7.8 Setup for measuring polarized photoluminescence.

There are numerous variations on this basic setup that can offer various improvements. For example, the quarter wave plates may be replaced by liquid crystal variable retarders that act the same as a wave plate, but with a retardation that can be set electronically. This removes the need to manually rotate polarizers or wave plates. Further, one of the quarter wave plates may be replaced by a photo-elastic modulator. A PEM is an oscillating variable retarder that can modulate the helicity of light between right and left circular polarization at typical frequencies of tens of kHz. The intensity of the light at the detector will then be modulated at this frequency, with the amplitude of oscillation proportional to the degree of polarization of the luminescence. A lock-in amplifier may then be used to detect this oscillating signal. If the helicity of the excitation light is modulated in this way, the build-up of nuclear spin polarization will also be suppressed.

Another modification to the setup shown in Figure 7.8 is to place optical filters in the collection path. By using high-pass, low-pass, or band-pass optical filters, a specific part of the emission spectrum may be selected for measurement. Additionally, if the optical excitation is nonresonant with the emitting energy levels, optical filters can be used to ensure that the excitation light does not reach the detector. In this case, if one uses filters in the collection path, the collection path may be placed in line with the excitation path, which is desirable in some situations.

Along similar lines, the detector shown in Figure 7.8 may be replaced by a spectrometer. Now, the entire spectrum of the emitted light may be recorded (by a CCD camera, for example), and the right and left circularly polarized spectra may be compared. This method is particularly useful in quantum dot studies where the luminescence from the dots must be measured separately from that of the wetting layer, or in single dot studies where luminescence lines from different dots or different transitions must be distinguished.

These methods for measuring polarized luminescence are useful for reading out the spin state of electrons and holes in quantum dots, however much more information can be obtained using time-resolved techniques to observe the dynamics of the spin states as they evolve in time.

7.2.1
Time-Resolved Photoluminescence

The experimental methods described above can be directly extended to measure luminescence as a function of time on picosecond or nanosecond timescales. There are at least three ways to accomplish this: (i) Photoluminescence up-conversion, (ii) Time-correlated photon counting, and (iii) A fast streak camera.

The first step in all time-resolved measurements is to initialize the system at a given time, $t = 0$. In all-optical experiments this is accomplished by using a pulsed laser for the excitation. For spin studies in quantum dots, a mode-locked Titanium-sapphire laser is often used. These lasers have tunable wavelength in the near-infrared range and output pulses with duration from about 100 fs to several ps. The repetition rate of the laser pulses is set by the cavity geometry and is typically around 80 MHz.

In Cortez *et al.* [230], discussed above, photoluminescence up-conversion is used to measure the temporal decay of the photoluminescence after the excitation (see, e. g., [248]). In this scheme, the luminescence and a time delayed probe pulse are overlapped and passed through a nonlinear material, typically β-barium borate (BBO), which emits "up-converted" photons at the sum frequency of two incoming photons, when phase matching conditions are met. Since this process is highly nonlinear in the incoming intensity, a change in the luminescence intensity coincident in time with the probe pulse provides a significant change in the intensity of the up-converted light. The probe light is provided by a pulse derived from the same pulse used for the excitation. By controlling the time delay between the excitation and the arrival of this additional pulse one can gate the measurement of the photoluminescence. This method can provide very high temporal resolution (only limited by the duration of the probe pulse, down to femtoseconds).

Time-correlated single photon counting (TCSPC) is a more direct way of time-gating a luminescence measurement. In this scheme, the excitation pulse is also sent to a fast photodiode. When this photodiode detects the photons of the pump pulse, it sends a "start" signal to a second photodiode. The luminescence is sent to this second photodiode, which begins waiting for a photon detection as soon as it receives the "start" signal. Once a photon from the luminescence is received, the second photodiode sends a "stop" pulse. The time delay between the "start" and "stop" is then recorded, and the sequence is repeated many times. By building up a histogram of these delay times, the temporal profile of the luminescence can be mapped out. This is a powerful technique that has many variations, though limited by the speed of the electronics.

One final method for measuring time resolved photoluminescence is by using a streak camera. In such a device, the luminescence is incident on a photomultiplier tube, which converts the stream of photons to a stream of electrons. The electrons then pass through a set of electrodes with a fast ramped voltage, synchronized to the pump pulse. This ramping voltage causes the electrons to spread out spatially (into a streak), with the electrons entering first deflected less than those that enter later. The electrons are then measured by a spatially-resolved detector where one end of the streak corresponds to photons that hit the photomultiplier tube at $t = 0$, and then proceeding linearly in time along the streak.

7.2.2
Spin Storage and Retrieval

The techniques discussed above for measuring time-resolved PL, along with a polarization measurement, provide a direct measure of the spin state of the recombining exciton or trion state. In the case of trion PL, the spin of the electron in the dot prior to trion formation may be inferred by further analysis. A more direct way of observing the time dependence of a single spin state in a quantum dot was developed in M. Kroutvar *et al.*, "Optically programmable electron spin memory using semiconductor quantum dots" [232]. In this work, a layer of self-assembled InAs quantum dots are embedded in an epitaxially grown device that allows for the control of the charge state of the dots with a bias voltage. Moreover, the device contains a p-doped region separated from the dots by a narrow undoped GaAs barrier. Under the appropriate bias conditions when electrons and holes are optically excited into the dots, the holes will rapidly tunnel out into the p-doped reservoir while the electrons remain in the dot. This allows for a means of injecting single, spin polarized electrons into the dots when the excitation is circularly polarized.

The spin of these electrons can then be read out a time later by applying a voltage pulse that injects holes from the reservoir back into the dots. This scheme for initializing and reading out spins is illustrated in Figure 7.9. In part 7.9a, circularly polarized light resonantly pumps spin polarized electrons and holes into the quantum dots. (At most one electron–hole pair can be injected into each dot, since additional interaction energy would be required to add any additional carriers.) In this initialization stage, a voltage is applied across the device to remove the hole from the dot and send it into the *p*-doped hole reservoir. For a time Δt (Figure 7.9b), the electron is stored in the dot. Then, the bias across the device is changed to send unpolarized holes from the reservoir back into the dots (Figure 7.9c). This hole may then recombine with the electrons, and the circular polarization of this electro-luminescence indicates the spin of the stored electrons. A single-photon counter (photodiode) is used to detect the luminescence after it has passed through

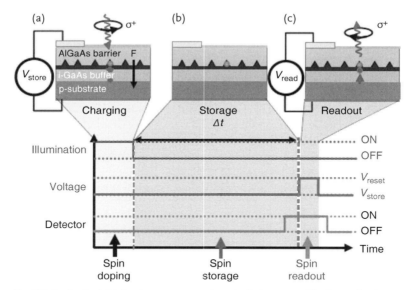

Fig. 7.9 Device for initialization, storage, and readout of spins in quantum dots. (a) Spin polarized electron–hole pairs are excited into the dots, and the hole quickly tunnels out. (b) The electron is stored in the dot for a time Δt. (c) A bias is applied to the device, causing the hole to tunnel back into the dot, resulting in luminescence, which is measured by a detector turned on only during this phase of the experiment. Reprinted by permission from Macmillan Publishers Ltd.: Nature [232], copyright (2004).

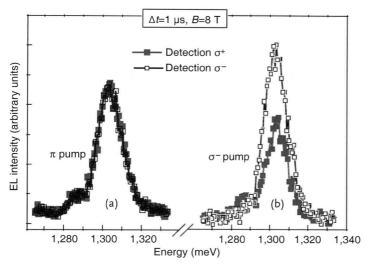

Fig. 7.10 Electroluminescence from the quantum dots after a storage time of 1 μs, in a magnetic field of 8 T. With a linearly polarized pump (a), the luminescence shows no circular polarization, but does with a circularly polarized pump (b). Reprinted by permission from Macmillan Publishers Ltd.: Nature [232], copyright (2004).

a spectrometer. To avoid detecting the excitation light, the detector is only turned on during the readout phase of the measurement.

This method provides the ability to initialize an ensemble of single electron spins, then store them for a set time, then read out the remaining spin polarization. In this work, the technique was employed to study the longitudinal electron spin relaxation time as a function of an applied magnetic field. Figure 7.10 shows the σ^+- and σ^--polarized readout electro-luminescence spectra 1 μs after the initialization pulse, and in an applied magnetic field of 8 T parallel to the spin polarization. On the left, the excitation is linearly (π-) polarized, and no net circular polarization is observed in the luminescence, indicating zero electron spin polarization. However, with a σ^- polarized pump, there is a clear σ^- circular polarization of the luminescence. This indicates that the electrons have retained their spin polarization after a storage time of 1 μs.

Figure 7.11a shows the circular polarization of the readout spectra as a function of storage time for both σ^+ and σ^- excitation. Here, the temperature is 1 K, and an 8 T longitudinal magnetic field is applied. One can clearly see that the σ^+-pumped spins yield σ^+ polarized excitation initially, decaying over time to the opposite polarization. On the other hand, the σ^--pumped spins yield the σ^- polarized PL with no appreciable decay in the observed time window.

These results can be understood in terms of the insets to Figure 7.11a. The two lines indicate the Zeeman-split spin states occupied by the electron in the quantum dot. σ^+-polarized excitation excites the electron into the upper sublevel. Over

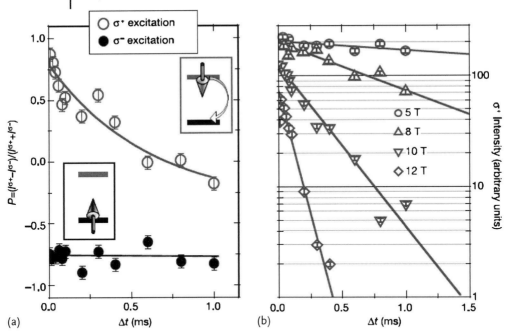

Fig. 7.11 (a) Spin memory as a function of storage time. Spins initialized in the higher-energy state thermalize to the lower energy state with a relaxation time on the order of ms. (b) Spin memory as a function of storage time at various magnetic fields (*y* axis is on a logarithmic scale). Reprinted by permission from Macmillan Publishers Ltd.: Nature [232], copyright (2004).

a timescale characterized by the spin relaxation time T_1, the electron may transition to the lower level. Under these experimental conditions, the thermal occupation of the upper state is quite low (about 2%), so the electron effectively relaxes to the lower level and then stays there. Likewise, the σ^- excitation injects the electron in to the lower energy sublevel, and therefore it just stays there.

By fitting the σ^+-excitation data to a decaying exponential, the spin relaxation time (T_1) can be extracted. Figure 7.11b shows the same type of data as in part a, at various magnetic fields. In this case, the *y* axis is plotted on a log scale, so the exponential dependence of the decay appears as a linear decrease.

At each magnetic field, the decay follows a single exponential, but the T_1 time decreases dramatically as the magnetic field is increased. In fact, the dependence of T_1 on magnetic field is found to obey a power law with an exponent of approximately -5. This dependence is consistent with a theoretical explanation for the spin flip mechanism in which acoustic phonons mediate spin-flips between the Zeeman levels by way of the spin–orbit interaction [249, 250].

The longest spin relaxation time measured in these experiments was $T_1 = 20$ ms at a magnetic field of $B = 4$ T and temperature $T = 1$ K. The detection sensitivity prevents measurement at lower magnetic field, but the T_1 time of these self-assembled quantum dots is likely to continue increasing as the field is lowered.

Alternatively, this device can be designed to have an electron-doped reservoir, which allows for the storage and readout of hole spin. This was demonstrated by D. Heiss *et al.* [251], and the hole spin T_1 time was found to go up to hundreds of microseconds. While not as long as the electron spin relaxation time, this is much longer than the unconfined hole spin lifetime. Given this relatively long lifetime, as well as the lack of coupling to nuclear spins mentioned above, it may be useful to use hole spins as qubits.

The measurements described above provide a measurement of the spin relaxation time, T_1, but are not able to see the spin coherence time, T_2. D. Heiss *et al.* [252] have proposed a modification to this experiment that would allow the observation of coherent spin dynamics in a similar setup. In this proposal, the idea is to perform the same optical electron spin initialization and storage in the diode device. But while the spin is stored, a second optical pulse re-excites the dot, spin dependently. That is, a circularly polarized pulse is used so that if the spin is polarized in one direction the trion state is generated, and if the spin is in the other state the trion cannot form due to Pauli blocking. After this second excitation the hole again rapidly tunnels out of the dot, leaving either one or two electrons in the dot, depending on the prior spin of the first electron. Now, when the readout phase arrives and a hole tunnels back into the dot, trion emission will be observed if the dot contained two electrons, but only exciton emission if the dot contained a single electron. By this difference in the emission, one can determine the spin of the electron at the time the second excitation pulse arrived. In between the first excitation pulse, coherent manipulation of the spin may potentially be performed, perhaps by electron spin resonance, or other optical means (see below).

Though this proposal has not yet been put into practice exactly as described, something similar has been demonstrated in the work of Ramsay *et al.* [253] Here, a similar scheme is employed with a hole being initialized and stored, except the final readout is performed electrically instead of optically via the luminescence. The experiment proceeds as described above, and the second excitation laser resonantly and spin-selectively drives the transition to the positively-charged trion state. This results in spin dependent Rabi oscillations between the single hole and trion state. If the second laser intensity is chosen to induce a Rabi oscillation through an angle of π, then the dot either contains a single hole, or a positively charged trion, depending on the initial hole state. So far, this is essentially the same as in the Heiss proposal. Now however, if the trion state has been formed, the electron will rapidly tunnel out of the dot (and later, the two holes will tunnel out as well). This results in a measurable current through the device, as was first demonstrated in the work of A. Zrenner *et al.* [42]. This photocurrent provides a sensitive measurement of the spin in the dot at the time of arrival of the second excitation pulse. In the present work, this allows time resolved measurement of the hole spin in the dot with few picosecond resolutions. One drawback to this work is that the hole spin being studied tunnels out of the dot within a few nanoseconds (as compared to the electrons which tunnel in about 50 ps). This limits the time that the spin can be observed. This may be overcome by a device of a different design, or perhaps by implementing the optical readout of the Heiss proposal above.

7.2.3
Magnetic Ions

Though not discussed at length in this book, magnetic atoms may be incorporated into quantum dots and their spin probed by optical means. This has been beautifully demonstrated in the work of Besombes *et al.* "Probing the spin state of a single magnetic ion in an individual quantum dot" [233]. This work focuses on single, MBE-grown CdTe quantum dots embedded in ZnTe. At the appropriate time during growth, manganese atoms are added to produce roughly equal quantum dot and manganese densities. Some of the dots will then contain a single manganese atom.

The spin of the manganese atom ($S = 5/2$) couples to the spin of a "bright" ($J = 1$) exciton in the quantum dot, resulting in splitting of the exciton states. The exchange interaction between the $2S + 1 = 6$ manganese angular momentum states results in a sixfold splitting of the exciton levels, as shown in Figure 7.12.

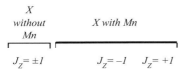

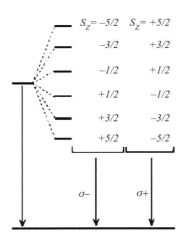

Fig. 7.12 Splitting of the exciton (X) state in a quantum dot due to interaction with a single manganese atom. The state may decay via emission of a σ^+ or a σ^- photon. Reprinted with permission from [233]. Copyright (2004) by the American Physical Society.

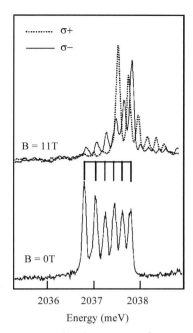

Fig. 7.13 Photoluminescence from a quantum dot containing a single manganese atom. At $B = 0\,T$, the spectrum is split into six more or less equal lines. At $B = 11\,T$, the lower-energy states of the manganese spin occur with higher probability. Reprinted with permission from [233]. Copyright (2004) by the American Physical Society.

This splitting can be directly observed as a sixfold splitting of the single dot PL spectrum (Figure 7.13) at zero magnetic field.

The interaction between the quantum dot and the manganese atom may be turned on and off by electrically tuning the charge state of the dot. Further, by performing photon correlation measurements on this sixfold emission, dynamics of the manganese spin can be investigated. Such measurements have found fluctuations in the Mn spin state with a characteristic time of about 20 ns. These fluctuations are attributed to interactions with the optically excited carriers. If these interactions could be suppressed, it is possible that the Mn spin state could be preserved for significantly longer [254].

The interaction of a Mn impurity spin and carrier spins in an InAs quantum dot has also been investigated in the work of A. Kudelski *et al.* [255] In this case, the interaction is somewhat more complicated. In the II–VI CdTe quantum dots discussed above, the Mn energy levels are isolated from the quantum dot levels, and the spins only interact through their exchange interaction. In InAs/GaAs dots, however, the Mn levels are such that the Mn defect forms an acceptor state with an activation energy of 113 meV. That is, the Mn atom has a hole bound to it with a binding energy of 113 meV. If the defect atom is ionized, a hole is added to the valence band of the semiconductor. By using photoluminescence spectroscopy of a single dot coupled to a Mn atom in a magnetic field, the interaction between the Mn ion spins, the bound hole spin, and optically excited quantum dot carriers is investigated.

7.2.4
Spin-Selective Absorption

In contrast to measuring emitted light, one can also probe spins in quantum dots using the absorption of light. The general scheme is illustrated in Figure 7.14. In a quantum dot containing a single electron in the spin-up state, transitions to the spin-down state may occur, but transitions to the spin-up state are forbidden by the Pauli exclusion principle. Thus, circularly polarized light of one helicity will not interact with the dot, while the other helicity may be absorbed or resonantly scattered.

This spin measurement technique was demonstrated in A. Högele *et al.*, "Spin-selective optical absorption of singly charged excitons in a quantum dot" [234]. Here, InGaAs self-assembled quantum dots are incorporated in a charge-tuning device as described in Chapter 4, and elsewhere. A single dot is located and characterized using photoluminescence spectroscopy. With the dot tuned into the singly-charged regime, a narrow linewidth continuous wave laser is focused onto the dot. The transmitted light is then collected on the other side of the dot. A differential absorption technique is used to enhance the signal. By modulating the bias across the device, the energy of the transition is modulated due to the quantum confined Stark effect. This allows one to measure a small relative change in absorption due to the quantum dot, eliminating background fluctuations. The signal may also be maximized by focusing the light as tightly as possible on the dot. By using a solid immersion lens, substantial signal can be obtained, even from a single quantum dot [256].

At zero magnetic field, a single dip in the optical absorption is seen as the laser energy is tuned across the energy of the negatively charged trion transition of the quantum dot (Figure 7.15a). This type of single dot absorption measurement provides high resolution spectroscopic information about the dot. Here, the transition energy is found to be 1.27087 eV, with a full width at half maximum of 2.3 μeV.

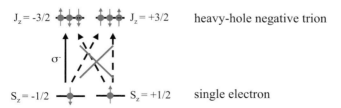

Fig. 7.14 Schematic of transitions from the single electron state to the negatively charged trion. For a given helicity of light (say, σ^-), only one transition can occur.

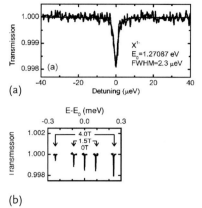

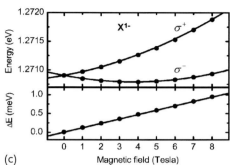

Fig. 7.15 (a) Transmission spectrum showing a dip due to absorption from a single electron-doped quantum dot. (b) Single-dot transmission spectra at 0, 1.5, and 4.0 T magnetic fields showing Zeeman splitting of the transition. (c) *Top:* Energy of the σ^+ and σ^-- polarized transmission dips as a function of applied magnetic field. *Bottom:* Splitting between the two sets of data above showing a linear Zeeman splitting. Reprinted with permission from [234]. Copyright (2005) by the American Institute of Physics.

When a magnetic field is applied, the electron energy level splits, resulting in a splitting of the observed transition energy into a doublet. As the field increases, the two levels will be occupied unequally as given by the thermal equilibrium (these experiments are carried out at $T = 1.5$ K). Figure 7.15b shows the absorption dip at $B = 0$ T, then the two Zeeman-split dips at $B = 1.5$ and 4.0 T. These spectra are recorded with linearly polarized light, and therefore are sensitive to both spin-polarized transitions. As the magnetic field is increased, one can see the increasing energy splitting, as well as a shift in the relative intensities of the two dips. This is attributed to the unequal thermal occupation of the two spin levels, yielding greater absorption in the less frequently occupied level.

The two different spin levels can be independently resolved by measuring the absorption of right or left circularly polarized light. Figure 7.15c shows the energy of the σ^+ and σ^- absorption dips as a function of magnetic field. As expected, the two levels split in energy linearly with magnetic field (shown in the bottom panel), with a parabolic overall shift due to the diamagnetic shift.

The experiment described above illustrates the general method of spin readout via absorption, but the technique is quite flexible and has been employed with a number of variations. The work above uses the absorption as a probe of the equilibrium spin population in the dot. In Atatüre *et al.* [161], discussed above, the same laser is responsible for pumping the spin state and also reading it out through the quenching of the absorption when the spin has been polarized. Of course, a second laser may be used as well to initialize the spin state before being probed by the change in spin dependent absorption of a second laser. For example, the Atatüre work was extended by M. Kroner *et al.* [257] to include a second laser, which allows independent spin pumping and readout revealing additional information about the physics involved. Another example of two-color spin preparation and absorption readout is in the work of D. Kim *et al.* [242], which was mentioned above. Here, one laser is used to pump a spin in a pair of coupled quantum dots, and the second laser is used to probe the spin of that electron using transitions involving the other quantum dot.

This method of spin readout via spin-selective absorption allows the measurement of spin states with high spectroscopic resolution. These types of techniques are reminiscent of atomic physics, and there are many analogies in terms of possible experiments. For example, the spin pumping discussed above in Atatüre *et al.*, and the Faraday rotation technique discussed in the next section are both closely related to techniques used in atomic physics experiments.

7.3
Observation of Spin Coherence and Optical Manipulation

Thus far, this chapter has focused mainly on initialization and measurement of spins into stationary eigenstates ("spin-up" or "spin-down"). Optical excitation can also be used to generate spin polarized electrons and holes in states that are not spin eigenstates. The resulting spin state will evolve in time, as introduced in Chapter 1. The same types of optical techniques discussed above can be applied to study these coherently evolving quantum states. Additionally, optical techniques may be employed to coherently control the spin state, as discussed in Sections 7.3.8 and 7.3.9.

7.3.1
The Hanle Effect

Using the polarization of time-averaged quantum dot luminescence as a probe of the spin polarization, one can obtain some information about the spin dynamics in a magnetic field. This type of experiment is known as a Hanle measurement [121], and the setup is essentially the same as for quantum dot luminescence measurements shown in Figure 7.8. A circularly polarized laser is incident on the sample in the +x direction, serving to inject spin-polarized electrons and holes. A magnetic field B_z is applied in the z direction. The polarization of the subsequent photoluminescence (PL) collected back along the x direction reveals the steady-state spin polarization along the measurement direction.

This steady-state spin polarization can be calculated by taking an initial electron spin polarization S_0 at $t = 0$ along the +x direction. (For simplicity, we will assume that the hole spin lifetime is very short and can be ignored.) The spin then precesses, and the x component of the spin as a function of time is given by

$$S_x(t) = \begin{cases} 0 & t < 0 \\ S_0 \cos(\omega_L t) \exp(-t/T_2^*) & t \geq 0 \end{cases}. \tag{7.2}$$

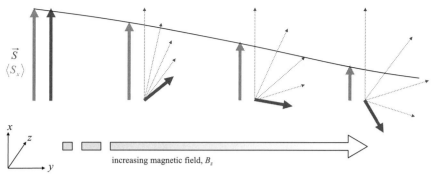

Fig. 7.16 Illustration of the Hanle effect. At zero magnetic field (on the left), spins initialized in the x direction do not precess and the time-averaged spin projected on the x direction is maximal. As the magnetic field is increased, the spin precesses through some angle within its lifetime, resulting in a reduced time-averaged projection along the x direction.

$\omega_L = g\mu_B B/\hbar$ is the Larmor precession frequency, and for generality both spin decoherence and dephasing are included in the spin lifetime T_2^*. The steady state spin polarization $\overline{S_x}$ is then found by integrating Eq. (7.2) from $t = (-\infty, \infty)$:

$$\overline{S_x} \propto \int_0^\infty S_0 \cos(\omega_L t) \exp(-t/T_2^*)\mathrm{d}t = S_0\left(\frac{1/T_2^*}{\omega_L^2 + (1/T_2^*)^2}\right). \tag{7.3}$$

Thus as a function of magnetic field (proportional to ω_L), the measured PL polarization sweeps out a Lorentzian function with width $B_{1/2} = \hbar/(g\mu_B T_2^*)$. This effect is illustrated in Figure 7.16. If the g-factor is known, then this measurement reveals the effective transverse spin lifetime, T_2^*. The analysis of the Hanle measurement can be done more rigorously and quantitatively by setting up and solving rate equations for spin injection, decay, and recombination [121]. The theory can be extended to the case of nuclear polarization, nonexponential decay, doped semiconductors, and so on This was the technique used for much of the initial exploration of spin physics in semiconductors, such as in the work described in *Optical Orientation*.

7.3.2
Ensemble Hanle Effect

The Hanle effect provides a straightforward way to obtain information about the spin coherence of electrons and holes in an ensemble of quantum dots. If one measures the photoluminescence polarization from an ensemble of quantum dots, the only necessary addition to perform a Hanle measurement is a transverse magnetic field. This was demonstrated, for example, by R. J. Epstein *et al.*, "Hanle effect measurements of spin lifetimes in InAs self-assembled quantum dots" [235]. In this work, undoped self-assembled InAs dots are placed in an optical cryostat with a superconducting magnet providing a transverse field (transverse to the light propagation direction). The setup is essentially the same as in Figure 7.8. Spin polarized electrons and holes are excited into the wetting layer (1.45 eV) using circularly polarized light. The emitted light is passed through a monochromator (or spectrometer) to measure the spin coherence time as a function of emission energy.

The circular polarization of the emission as a function of magnetic field (shown in Figure 7.17a) displays the expected peaked lineshape around zero field. However, the central peak does not decay all the way to zero, and in fact, a second, broader peak must be included in the fit to match the data. As described above, the width of the Hanle peak is related to the effective transverse spin lifetime by $B_{1/2} = \hbar/(g\,\mu_B\,T_2^*)$. Therefore, each Lorentzian component of the fit can be associated with a value of $g T_2^*$. The g-factor of these quantum dots was not measured, but similar dots have been found to have $g \approx 1.7$. Using this value, one can arrive at an approximate value for the effective spin coherence time, T_2^*.

The extracted values of $g T_2^*$ are shown in Figure 7.17b for three different samples as a function of magnetic field. Both the narrow and broad fit components are shown. The narrow peak yields a coherence time on the order of 100 ps at $T = 4$ K, decreasing as the temperature increases. The broad component yields a much shorter spin lifetime; on the order of a few picoseconds. The origin of this short spin lifetime component is unclear. One possibility may be the short-lived hole spin coherence.

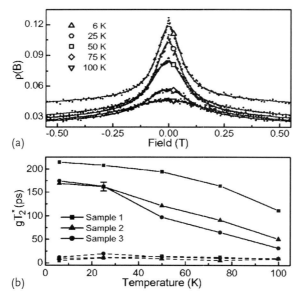

Fig. 7.17 (a) Hanle curves from InAs quantum dots at various temperatures. (b) $g T_2^*$ extracted from Hanle measurements as a function of temperature. Hanle data is fit to two lifetimes, shown as the solid and dashed data. Three samples are measured with different dot shapes (see [235] for details.). Reprinted with permission from [235]. Copyright (2001) by the American Institute of Physics.

Since the quantum dots in this experiment are undoped, optical excitation will tend to generate one or more electron–hole pairs in the dots. Therefore, the observed spin lifetime cannot be longer than the recombination time of an exciton in the dot. It is possible that this is the limiting factor in these experiments.

On close observation of the data in Figure 7.17a, one can see small, very sharp peaks around zero magnetic field in the lower temperature data. The origin of these peaks is not explained in this work, though it may stem from another, much longer spin lifetime component. One possibility is that it arises from the occasional generation of a trion state in some of the dots. As we will see in the next section, Hanle data from a negatively charged exciton state can reflect the single spin coherence time, independent of the electron–hole recombination time [211, 258]. In the present experiment, however, the neutral and charged exciton behavior cannot be easily disentangled since the emission energy of these two states differs by only about 4 meV – much less than the broadening of the PL peaks due to inhomogeneity within the ensemble. This inhomogeneity may be removed by looking at a single quantum dot, yielding much greater spectroscopic resolution. Clearly, there are a number of questions that cannot be answered by an ensemble Hanle measurement alone. For a more detailed understanding of spin coherence in quantum dots, we must turn to single quantum dot studies, and/or time-resolved experiments.

7.3.3
Hanle Effect Measurement of a Single Quantum Dot

In general, optical measurements of a single quantum dot provide additional information over ensemble measurements. This is illustrated in the previous section, where the spectroscopic resolution is not high enough to separate different transitions within the dots. Single dot measurements are typically more technically difficult, but can overcome some of the limitations of inhomogeneous broadening.

In the work of A. S. Bracker *et al.*, "Optical pumping of the electronic and nuclear spin of single charge-tunable quantum dots" [211], Hanle measurements are performed similar to those above, but looking at a single quantum dot.

The dots in these experiments are interface fluctuation quantum dots, formed in a 4.2-nm wide GaAs quantum well, with AlGaAs barriers. As has been discussed numerous times above in the case of self-assembled dots, these quantum dots are also embedded in a diode structure that allows for control over the charge state of the dots. Electrons and holes are excited into the quantum well (analogous to the wetting layer in self-assembled dots) with a circularly polarized laser. To isolate a single quantum dot, a metal mask layer is deposited on the surface of the sample with submicron apertures. The laser is focused onto the sample using a microscope objective, and the each aperture is checked to find one with a good-looking quantum dot directly underneath. This quantum dot is then characterized with photoluminescence spectroscopy.

Photoluminescence of the quantum dot as a function of the bias voltage across the device is shown in Figure 7.18a. Three different lines are identified in the spec-

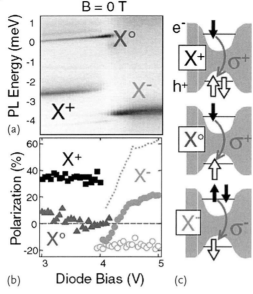

Fig. 7.18 (a) Photoluminescence from a single quantum dot as a function of applied bias. The neutral exciton (X^0), negatively charged exciton (X^-) and positively charged exciton (X^+) lines are indicated. (b) Circular polarization of the three lines indicated in (a). The different curves for the X^- state correspond to different excitation intensity or power (see [211]). (c) Illustration of the different recombination processes giving rise to the observed lines. Reprinted with permission from [211]. Copyright (2005) by the American Physical Society.

tra: the neutral exciton, X^0, the negatively charged exciton, X^{1-} (labeled X^-), and the positively charged exciton X^{1+} (labeled X^+). There is an abrupt change between the X^0 and X^+ lines and the X^- line around approximately 4 V. It is here that the equilibrium state of the dot switches from zero to one electron. When the dot is empty in equilibrium, optical excitation can generate the X^0 state, or if an excess hole is captured in the dot, the X^+ state. However, with an electron already resident in the dot, the X^- state is predominantly formed.

Figure 7.18b shows the degree of circular polarization of the three identified lines indicated in part a. The X^0 emission is almost entirely unpolarized, due to the anisotropic exchange interaction arising from the elongated shape of these quantum dots (see Chapter 2). The X^+ emission has a large, positive polarization. This state is composed of two holes in a singlet state, with the net spin polarization determined only by the electron. The positive polarization of this emission reflects the spin of the optically injected electrons. The polarization of the X^- emission is more complicated. The three curves shown in the figure correspond to different excitation energies or intensities. In some cases, the polarization is negative, similar to the case of self-assembled dots discussed above by Cortez et al. [230]. However, in this type of quantum dot other processes contribute that depend on the excitation energy, and the contribution from electrically injected electrons, which can lead to positive polarization of the X^- emission [211]. The diagrams in Figure 7.18c illustrate the different transitions yielding the observed emission.

The Hanle measurement is performed by monitoring the polarization of these different luminescence lines while sweeping an applied magnetic field perpendicular to the spin polarization axis. This is illustrated in Figure 7.19b, and the results are shown in part a. As expected, the X^+ and X^- lines are well-fit by a Lorentzian lineshape, while the X^0 polarization remains at zero.

From these curves, the spin coherence time T_2^* can be extracted from the width of the Hanle curves using the relation $B_{1/2} = \hbar/(g\,\mu_B\,T_2^*)$, and estimating $g \approx 0.2$ from previous measurements on these quantum dots. In this way, coherence times of 150 ps and 16 ns are found for the X^+ and X^- lines, respectively. The 150 ps line is consistent with the radiative recombination time in these quantum dots (and also consistent with the spin lifetime measured in ensembles of self-assembled dots described above). Clearly, the coherence time measured for the X^- emission is much longer than the recombination time.

The Hanle curve from the X^- emission is attributed to the behavior of the single electron in the quantum dot before the charged exciton (trion) formation. That is to say, the polarization of the X^- emission depends on the spin of the resident electron in the dot, before an additional electron and hole are added via optical excitation. This single electron spin coherence time is consistent with what is the expected dephasing of the spins due to interaction with the randomly oriented nuclear spin polarization (see Chapter 6). Other mechanisms, however, cannot be ruled out.

Although this measurement looks at a single quantum dot, it is still averaged over many repeated initializations of that quantum dot. Therefore, effects due to an inhomogeneous distribution of quantum dots are eliminated, but time-varying in-

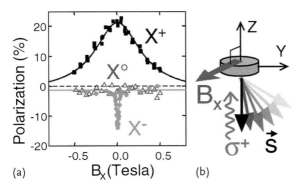

(a)   (b)

Fig. 7.19 (a) Hanle curves from the three transitions observed in the quantum dot. (b) Schematic of the experiment. Circularly polarized light excites the spin along the z direction, which then precesses around the magnetic field applied in the x direction. The steady-state spin along the z direction is measured via photoluminescence. Reprinted with permission from [211]. Copyright (2005) by the American Physical Society.

homogeneities remain. Most notably, the random nuclear spin polarization varies slowly in time, and may still affect the spin measured in these experiments. Still, these single dot measurements greatly increase the spectroscopic resolution, as well as removing the distribution of a quantum dot ensemble, which could lead to additional dephasing of the spin precession. To date, no optical measurements of spin in one or more quantum dots has been performed with a single initialization and readout cycle (a so-called single shot measurement). Some of these dephasing effects can be corrected by performing a spin echo-type measurement, which will be described below (see Section 7.3.5).

7.3.4
Time-Resolved Faraday Rotation Spectroscopy

The Faraday and Kerr rotation effects provide a very useful technique for observing the coherent spin dynamics of electrons in both bulk semiconductor systems as well as quantum dots. For one example, we will look at the results of J. Gupta *et al.*, in "Spin dynamics in semiconductor quantum dots" [236], which focuses on time resolved measurements of spin coherence in ensembles of nanocrystal quantum dots.

This work uses Faraday rotation measurements to observe the dynamics of optically excited electron and hole spins in CdSe nanocrystals. These nanocrystals are chemically synthesized as described in Chapter 2, and then suspended in a polymer film for measurements. The measurements are carried out in a magneto-optical cryostat at temperatures between 4 K and room temperature, with the ability to apply a magnetic field in the transverse direction.

When linearly polarized light is transmitted through matter with different refractive indices for left- and right-handed circularly polarized light, n_+ and n_-, a rotation of the plane of polarization is generated. Assume the light is of frequency ω and travels a distance L through matter. Due to different phase velocities of the two circularly polarized components, $v_\pm = \omega/k_\pm = c/n_\pm$, a phase difference $L(k_- - k_+) = (\omega L/c)(n_- - n_+)$ accumulates between them when passing through the sample. It can be shown geometrically that the rotation angle θ of the polarization plane of linearly polarized light is half the phase difference of the circular components. We therefore obtain

$$\theta = \frac{EL}{2\hbar c}(n_- - n_+) \,, \tag{7.4}$$

where $E = \hbar\omega$. This rotation of the plane of linearly polarized light is known as Faraday rotation. An analogous effect occurs when light reflects off of the surface – this effect is known as the Kerr rotation. As can be seen in the above equation, Faraday rotation (or Kerr rotation) is not related to extinction, it is only related to a relative change in the refractive indices n_- and n_+, which is generated due to magnetically polarized carriers. Thus, Faraday rotation can be regarded as a "magnetically induced" optical activity or optical rotation. The origins of Faraday and

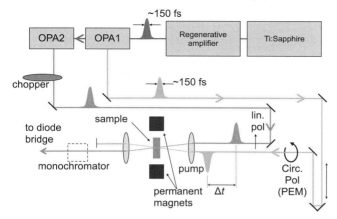

Fig. 7.20 Setup for time resolved Faraday rotation.

Kerr rotation will be discussed in a more quantum mechanical framework in Section 7.3.6.

The setup used for these measurements is illustrated in Figure 7.20. An amplified Ti:Sapphire laser is used to drive an optical parametric amplifier (OPA), which can produce pulses of light approximately 100 fs in duration, tunable in wavelength throughout the visible spectrum. This is the relevant wavelength range for CdSe structures, as well as quantum dots composed of other II–VI semiconductors. A second parametric amplifier may be synchronized to the first, allowing for pump and probe beams of differing wavelengths. Alternatively, the train of laser pulses may be divided into two paths by means of a beamsplitter to form pump and probe paths. The pump beam is reflected off of a translatable mirror before both beams are focused through the same lens onto the sample. By translating the mirror in the pump path, the distance traveled by the pump pulses can be adjusted relative to that traveled by the probe pulses. This allows fine control over the relative arrival time of the two pulses at the sample. Given the speed of light $c = 3 \times 10^8$ m/s, changing the position of the mirror in the pump path by 1 mm changes the arrival time at the sample by about 6.7 ps.

Before arriving at the sample, the pump beam is circularly polarized to excite spin polarized electrons and holes, and the probe beam is linearly polarized for the measurement of Faraday rotation. After passing through the sample, the pump beam is blocked and the probe beam, whose polarization has now been rotated due to the Faraday effect, is directed into the collection path.

Typical polarization rotations in such measurements tend to be on the order of 1 mrad or less. To achieve sensitive polarization measurements, a balanced photodiode bridge technique is often employed. As shown in Figure 7.21, the polarization of the probe beam, initially vertical, is rotated through an angle θ_F. The probe beam is then passed through a half-wave plate, which reflects the polarization about its axis at 22.5°, resulting in the polarization at an angle $\theta = \pi/4 - \theta_F$. Next, the probe is passed through a polarizing beamsplitter, which separates the horizon-

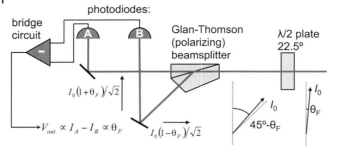

Fig. 7.21 Measurement of linear polarization rotation using a balanced diode bridge. Incoming light of intensity I_0 with polarization rotated from the vertical by a small angle θ_F is measured as an output voltage V_{out} proportional to θ_F.

tally and vertically polarized components of the light. Assuming θ_F is small, the horizontal and vertical components have intensity $I_{H(V)} \approx I_0(1 \pm \theta_F)/\sqrt{2}$ where the incoming probe beam has intensity I_0. These two beams are then focused onto a pair of photodiodes, and the difference between the two photocurrents is measured by the diode bridge circuit. This circuit contains a pair of photodiodes and at least one operational amplifier to take the difference of the two diodes. Additional operational amplifiers may be used to also output the signal from each photodiode. This difference signal $S_{H-V} \propto I_H - I_V = \sqrt{2}I_0\theta_F$, and is thus proportional to the Faraday rotation angle.

The signal-to-noise is improved by employing lock-in amplifier techniques. The pump beam is modulated between left and right circular polarization using a photo-elastic modulator (PEM) at a frequency of tens of kHz, and the probe beam is modulated (turned on and off) at a frequency of 100s of Hz using a mechanical chopper (a rotating wheel with alternating opaque and open regions). The output of the photodiode bridge is then sent to a lock-in amplifier referenced to the PEM, and the output of this lock-in is sent to a second lock-in, referenced to the chopper frequency. This has the effect of eliminating various noise sources, as well as only measuring the component of the probe light modified by the injected spins.

Typical time-resolved Faraday rotation traces are shown in Figure 7.22a, in a transverse magnetic field of 0.25 T, and at a temperature of 6 K. Here, samples of CdSe nanocrystals, as well as CdSe cores overgrown with a thin cap of CdS are measured. At $t = 0$, the pump pulse arrives and excites typically at most one electron–hole pair in a given quantum dot. Since the initial state is in a superposition of the spin-up and spin-down eigenstates (as defined by the applied field), the spin state then begins to precess. When the probe pulse arrives, it projects the state onto the laser propagation direction, yielding a cosinusoidal trace as a function of the probe delay. Additionally, the signal decays with time due to spin decoherence and dephasing. From the frequency of the observed oscillations, one can extract the g-factor for spins in the quantum dots, and the decay of the signal provides a measure of the effective spin coherence time T_2^*. Note that the spin decoherence here is not limited by electron–hole recombination, since the radiative lifetime is

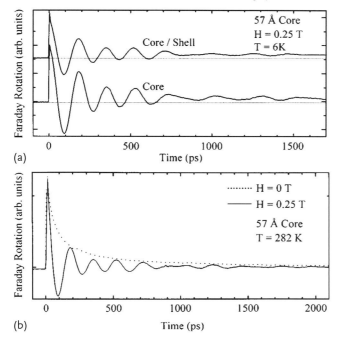

Fig. 7.22 (a) Time-resolved Faraday rotation data from CdSe quantum dots, and CdSe dots capped with a CdS shell. The data show the projection of the spins as they coherently precess around a magnetic field. (b) Coherent spin precession near room temperature. The dotted line shows the spin dynamics in zero magnetic field. Reprinted with permission from [236]. Copyright (1999) by the American Physical Society.

very long in these dots (see Chapter 2). At higher magnetic field, the decay of the spin precession is due to dephasing due to the distribution of quantum dot size and shape. Since each nanocrystal has a slightly different g-factor, the spins become out of phase with each other leading to a cancelation of the total signal. Apart from such dephasing it is not well known what the dominant decoherence mechanisms are in such quantum dots. Interaction with randomly oriented nuclear spins has been suggested to give the right order of magnitude [226], but little experimental evidence exists.

Figure 7.22b shows the coherent spin precession in these CdSe quantum dots near room temperature ($T = 282$ K). Remarkably, there is essentially no difference from the data at $T = 6$ K. The strong confinement of electrons in these quantum dots provides large level splittings, allowing the nanocrystals to continue functioning as quantum dots even at room temperature.

The data in Figure 7.22 shows some slight deviations from perfect single-frequency oscillatory behavior. This can be seen more clearly at higher applied fields (Figure 7.23a). At 4.0 T, the Faraday rotation trace shows clear beating behav-

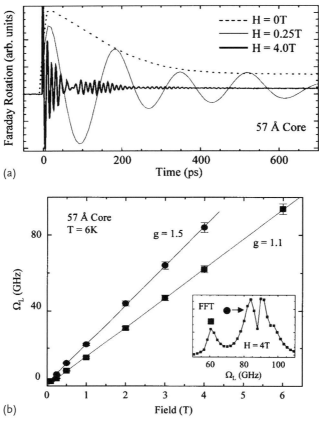

Fig. 7.23 (a) As the magnetic field is increased, the lifetime measured by time-resolved Faraday rotation becomes shorter due to inhomogeneous dephasing. At high magnetic field, the presence of multiple precession frequencies becomes clear. (b) Two of the measured precession frequencies as a function of magnetic field from an ensemble of CdSe quantum dots. Inset: Fourier transform of a time-resolved trace showing multiple precession frequencies. Reprinted with permission from [236]. Copyright (1999) by the American Physical Society.

ior, due to precession at multiple distinct frequencies. These precession frequencies are found to vary linearly with the magnetic field (Figure 7.23b), corresponding to multiple distinct *g*-factors. The observation of multiple *g*-factors is ubiquitous in studies of spin precession in nanocrystal quantum dots. Evidence suggests that these different precession frequencies are related to electron vs. exciton spin precession [259], or the state of charge traps on the surface of the dots [115]. Further study is needed to fully understand the dynamics of spin states in these systems.

Despite the mystery of multiple *g*-factors, all of the *g*-factors show an expected monotonic shift with nanocrystal size. Figure 7.24 shows the measured *g*-factors (open symbols) as a function of CdSe nanocrystal diameter. The lines correspond to theoretical calculations. The lowest *g*-factor agrees with a theoretical description

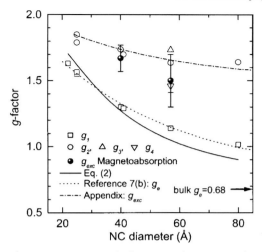

Fig. 7.24 Two measured *g*-factors in CdSe quantum dots, as a function of nanocrystal diameter. Solid points indicate *g*-factors measured by magnetoabsorption. Lines show different theoretical models (see [259]). Reprinted with permission from [259]. Copyright (2002) by the American Physical Society.

in terms of the *g*-factor shift for single electron spins with increasing confinement (see Chapter 3). The higher *g*-factors agree with a calculation of the precession of electron-hole pairs, described in [259].

7.3.5
Coherent Spin Echos – Measurement of T_2

In all optical studies of spins in quantum dots to date, averaging over multiple iterations is performed either in time with a single dot, or both in time and over an ensemble of dots. The ability to perform optical single-shot readout of a single spin in a quantum dot would be a landmark achievement. As is seen in nuclear spin resonance experiments, however, some information lost from ensemble averaging can be regained by spin echo techniques [225]. This idea has been most strikingly employed in the area of optically active spins in quantum dots in the work of A. Greilich *et al.*, in "Mode-locking of electron spin coherences in singly charged quantum dots" [237] and "Nuclei-induced frequency focusing of electron spin coherence" [245].

In traditional nuclear spin echo techniques, short radio-frequency pulses are applied to coherently manipulate the spins through well defined angles around the Bloch sphere. The same RF technique may be applied to spins in quantum dots [260, 261], though the relatively short coherence time of electron spins compared to nuclear spins limits the practicality of this technique. The ability to perform fast coherent spin rotations using optical pulses is now becoming possible in quantum dot systems (see Sections 7.3.8 and 7.3.9), though spin echo measurements have not yet been performed along these lines. Instead, A. Greilich *et al.* have taken advantage of a surprising effect in which the precession frequencies in an ensemble of quantum dots automatically synchronize in a way that leads to the rephasing of the spins.

The samples in these experiments consist of multiple layers of InGaAs self-assembled quantum dots, with delta-doped layers included to put, on average, one electron in each dot. Optical pump pulses are provided by a mode-locked Ti:Sapphire laser emitting pulses of 1.5 ps duration, once every 13.2 ns. These pump pulses are circularly polarized and resonantly excite trions which, on recombination, leave behind a spin polarized electron. The recombination time of the trion is a few hundred picoseconds, which is seen only as a distortion of the signal on these short timescales.

The subsequent dynamics of the electron spins are monitored through time-resolved Faraday rotation, as in the previous section. Figure 7.25 shows the measured spin precession at three different magnetic fields. At positive times (with the Faraday rotation probe pulse arriving after the pump), typical spin precession is observed with a decay time that decreases with increasing magnetic field. This is similar to the dephasing discussed in the previous section, attributed to the inhomogeneous range of g-factors among the ensemble of quantum dots. This dephasing causes the signal to disappear within a few nanoseconds though, crucially, the coherence of individual spin states may persevere despite being out of phase with the ensemble. Interestingly, unlike in the previous section a nearly symmetric sig-

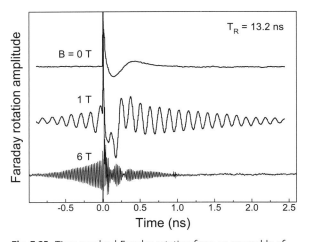

Fig. 7.25 Time-resolved Faraday rotation from an ensemble of singly-charged InAs quantum dots. The signal decays at positive time due to inhomogeneous dephasing and then comes back into phase at negative times due to frequency-locking of the spin precession. Complex behavior near zero delay is attributed to the presence of the trion state. Figure courtesy A. Greilich, D. R. Yakovlev and M. Bayer, TU Dortmund.

nal is seen at negative times, with the probe pulse arriving just before the pump pulse.

To understand these results, it is important to bear in mind that the pulses are repeated every 13.2 ns. Thus, for example, -500 ps in the figure corresponds to the probe arriving 12.7 ns *after* the previous pump pulse. The interpretation of the data is that the repeated excitation somehow takes the continuous distribution of precession frequencies, and concentrates them into certain discrete values. Specifically, the only allowed spin precession frequencies ω_e are those that satisfy $T_R = 2\pi N/\omega_e$, where T_R is the time between pump pulses and N is an integer. That is, a whole number of spin precessions must fit evenly into the repetition period of the laser. With this condition, the ensemble of spins dephases as usual, but at a multiple of the repetition period all of the spins come back into phase with each other.

The explanation for this phenomenon was given in [245]. The key to this effect is that when the trion state is optically excited by the pump pulse, electron-nuclear spin flips that are otherwise forbidden may occur. Then first consider a quantum dot whose precession frequency already satisfies the condition above. After an electron spin is initially polarized (say in the x direction), at any multiple of 13.2 ns later the subsequent pump pulses will be unable to re-excite the dot since it will already contain a spin in the x direction. That is, when the pump pulse arrives, a trion cannot form due to the Pauli exclusion principle. Since trion formation or decay is needed for the electron-nuclear spin flips to occur, the nuclear spins around this dot will be frozen, at least on timescales shorter than the natural nuclear spin fluctuations. On the other hand, consider a quantum dot precessing at a frequency not satisfying the condition above. When subsequent pump pulses arrive, the spin will have at least some projection onto the negative x direction. Thus a trion may be excited in such a dot, re-initializing the electron spin as well as allowing for a perturbation to the nuclear spin polarization.

With a continuous train of optical pumping, quantum dots whose precession frequencies do not match up with the repetition period undergo repeated perturbations of their surrounding nuclear spin polarization. Eventually, the nuclear spins will be polarized such that they generate an effective magnetic field exactly bringing the precession frequency into one of the resonance frequencies. Once this occurs, this dot will join the subset of dots with frozen nuclear spin. Over time, virtually all of the dots in the ensemble will have achieved this frequency-locked state. In fact, this phenomenon shows remarkable memory effects over timescales of minutes due to the long spin memory of the nuclei.

This spin echo effect provides a nice way of measuring the electron spin coherence time, T_2, free from inhomogeneous dephasing effects. The Faraday rotation signal observed at "negative" time comes from contributions to the signal from spins initially polarized at 13.2 ns, 26.4 ns, and so on, earlier. Therefore, the signal will have decayed somewhat compared to the initial values immediately after $t = 0$. If the laser pulse repetition time is increased, this negative time signal should become smaller in accordance with the spin coherence time.

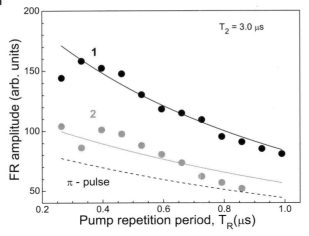

Fig. 7.26 Amplitude of the negative-time Faraday rotation signal as a function of the pulse repetition period. Curves 1 and 2 are for two different pump intensities. Fits to these data yield $T_2 = 3.0\,\mu s$. Figure courtesy A. Greilich, D. R. Yakovlev and M. Bayer, TU Dortmund.

Figure 7.26 shows the magnitude of the negative time data as a function of the time between pump pulses. The solid lines compare this data to the theory, which has the coherence time T_2 as its only fitting parameter. From this theoretical fit, the T_2 time is extracted and found to be $T_2 = 3.0\,\mu s$. This coherence time is three orders of magnitude longer than the dephasing times observed without the spin echo. It is also two orders of magnitude longer than spin lifetimes measured in single dot studies that include a time average over many iterations [211, 238]. This remarkably long T_2 time provides motivation that some application may be found for these systems in the area of quantum information processing.

7.3.6
Single Spin Kerr Rotation Measurement

The Faraday rotation measurements discussed thus far have all been performed on large ensembles of spins. Such measurements on single quantum dots are also possible. Although the signal from a single dot is smaller than from an ensemble, several factors work in one's favor. The laser can be focused as tightly as possible on the dot. Also, the Faraday or Kerr rotation signal from a single dot is concentrated spectrally in a smaller region, whereas an inhomogeneous ensemble it is spread over a larger range in energy. By using a narrow linewidth laser, one can look only in the relevant spectral range around the dot resonance. A cavity may also be used to enhance the optical response of the dot, increasing the single spin signal.

The work of J. Berezovsky *et al.*, "Nondestructive optical measurements of a single electron spin in a quantum dot" [60] demonstrates the ability to observe the electron spin in a single quantum dot using Kerr rotation spectroscopy. These re-

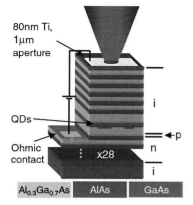

80nm Ti, 1μm aperture

i

QDs

p

Ohmic contact   x28   n

i

Al$_{0.3}$Ga$_{0.7}$As   AlAs   GaAs

Fig. 7.27 Schematic of the sample for single spin measurements. The sample consists of a layer of GaAs interface fluctuation quantum dots surrounded by distributed Bragg reflectors forming an optical cavity. The structure is doped and gated so as to control the charging of the dots and the quantum well.

sults present a continuously time-averaged measurement. This technique can be extended into the time domain, as described in Section 7.3.7.

A schematic of the sample used in this work is shown in Figure 7.27. The sample is grown by molecular beam epitaxy and consists of a single 4.2-nm GaAs quantum well in the center of a planar Al$_{0.3}$Ga$_{0.7}$As λ-cavity. A 2-min. growth interruption at each quantum well interface allows interface fluctuation quantum dots to form.

The quantum dot layer is centered within an optical microcavity with a resonance chosen to enhance the interaction of the optical field with the quantum dot. See Section 4.2 for more discussion of cavity-enhanced Faraday rotation. The front and back cavity mirrors are distributed Bragg reflectors composed of 5 and 28 pairs of AlAs/Al$_{0.3}$Ga$_{0.7}$As $\lambda/4$ layers, respectively. This asymmetrical design allows light to be injected into and emitted from the cavity on the same side. Based on previous measurements with similar cavities [175, 262], the Kerr rotation enhancement is expected to be enhanced by a factor of ~ 15 at the peak of the cavity resonance.

Additionally, the quantum dots are embedded in a diode-like structure, similar to those discussed previously, allowing the charging of the dots and the quantum well to be controlled with a bias voltage. On the top surface of the device, a metal layer forms a Schottky contact with 1-μm apertures fabricated by electron-beam lithography. This layer serves as both a front gate and a shadow mask for isolating single dots.

As discussed in Section 7.3.4, the magneto-optical Kerr effect results in a rotation of the plane of polarization of linearly polarized light with energy E upon reflection off the sample, and is analogous to the Faraday effect for transmitted light. For both effects, the rotation angle is determined by the difference of the dynamic dielectric response functions for σ^+ and σ^- circularly polarized light, which are proportional to the interband momentum matrix elements, $P_{c,v}^\pm = \langle \psi_c | \hat{p}_x \pm i\hat{p}_y | \psi_v \rangle$, where ψ_c (ψ_v) is a conduction (valence) band state [134, 263]. Due to the cavity, both reflection and transmission contribute to the measured polarization rotation. For simplicity, we refer only to Kerr rotation. For a single conduction-band energy level in a quantum dot containing a spin-up electron in a state $\psi_\uparrow$, optical transitions to the spin-up state are Pauli-blocked, and the KR angle is then given by

$$\theta_K(E) = C E \sum_\nu \left(|P_{\downarrow,\nu}^+|^2 - |P_{\downarrow,\nu}^-|^2 \right) \frac{E - E_{0,\nu}}{(E - E_{0,\nu})^2 + \Gamma_\nu^2} , \qquad (7.5)$$

where C is a constant, and $E_{0,\nu}$ and Γ_ν are the energy and linewidth of the transition involving the valence band state $|\psi_\nu\rangle$, respectively. For a single transition in the sum, with $\Gamma \ll |\Delta| \ll E$, where $\Delta = E - E_0$, note that $\theta_K \sim \Delta^{-1}$, which decays more slowly than the absorption line, ($\sim \Delta^{-2}$) [134, 264]. Therefore, for a suitable detuning, Δ, KR can be detected while photon absorption is strongly suppressed. This gives the Kerr rotation measurement a nondestructive property, whereby the spin may be minimally affected by the measurement.

A continuous wave, circularly polarized Ti:Sapphire laser is used to excite spin-polarized electrons and holes. A second, linearly polarized continuous wave Ti:Sapphire laser is used as the probe for Kerr rotation measurements. The two beams are made collinear and sent into the same microscope objective. The objective focuses both beams onto the sample (spot size $\sim$1 μm), which sits inside a liquid Helium flow cryostat. Additionally, the iron poles of an electromagnet are positioned above and below the sample cold finger to apply magnetic fields up to $\sim$0.1 T.

The light reflected off (or emitted from) the sample is collected through the objective and passed through a long-pass optical filter to block the pump beam. The rotation of the probe polarization is then detected by a balanced photodiode bridge, as described in Section 7.3.4. The difference channel of the diode bridge is sent to a voltage preamplifier and two lock-in amplifiers, and the signal is averaged for several seconds to reduce noise. During this measurement time of several seconds, the pump repeatedly re-initializes the spin. In this sense, it is a measurement of a single spin in a dot repeated many times and averaged in time. Finally, the pump polarization is switched between σ^+ and σ^- with a liquid crystal variable waveplate, and a measurement of the rotation angle is taken at each helicity. The difference between these two values yields the signal modulated at both the pump and probe frequencies, and which depends on the sign of the pump helicity. Sweeping the probe laser energy with a measurement at each energy maps out the spectral dependence of the Kerr rotation.

First, a good-looking QD is found and characterized using photoluminescence spectroscopy. Figure 7.28 shows the luminescence as a function of bias voltage for a single quantum dot. Also shown is the polarization of the emission of two of the observed lines. This is similar to the behavior described above in Section 7.3.3. In this case, three different lines are identified. Above 0.5 V a single line is observed at 1.6297 eV, which is attributed to recombination from the negatively-charged exciton (trion, or X$^-$) state. Below 0.5 V this line persists faintly, and a bright line appears 3.6 meV higher in energy due to the neutral exciton (X^0) transition. The presence of the X$^-$ line at $V_b < 0.5$ V implies that occasionally a single electron is trapped in

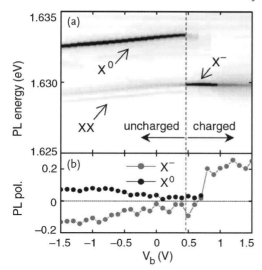

Fig. 7.28 (a) Single dot photoluminescence versus bias voltage. The neutral exciton (X^0), charged exciton (X^-) and biexciton (XX) lines are identified. (b) The degree of circular polarization of the X^0 (black) and X^- (gray) PL lines as a function of bias voltage. The biexciton PL is unpolarized.

the dot, forming an X^- when binding to an electron and a hole. In addition, a faint line at 1.6292 eV is visible from radiative decay of the biexciton (XX).

The X^0 PL shows a small, but positive polarization over the entire range of bias where X^0 PL is present. This reflects the polarization of the injected electrons and holes. The magnitude of the circular polarization is most likely reduced from the polarization of the injected carriers due to the anisotropy of the electron-hole exchange interaction in the dot. This effect arises from elongation of the dots along the $[\bar{1}10]$ crystal axis [24], causing electrons and holes to relax into states emitting linearly polarized light.

The polarization of the X^- line is similar to that observed in Section 7.3.3. After the recombination of the X^-, the negative circular polarization seen over much of the bias range means that the electron left in the dot is polarized in the spin-up direction. (For the purposes of this discussion, "spin-up" will refer to the polarization direction of the optically excited spins.) In this way, both optical injection and trion recombination serve to pump lone spin-up electrons into the dot.

The data in the top panel of Figure 7.29 show the Kerr rotation signal as a function of probe energy for σ^+ and σ^- pump helicity. Here, the applied bias is $V_b = 0.2$ V and the quantum dot is in the uncharged regime. The photoluminescence at this bias is also shown, with the X^- and X^0 energies indicated by the dotted lines. These energies coincide spectrally with two sharp features observed in the Kerr rotation data, labeled Ξ^- and Ξ^0, respectively. In the bottom two panels of Figure 7.29 the sum and difference of the σ^+ and σ^- data is shown. The feature

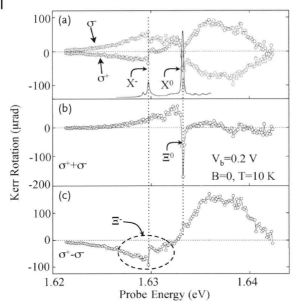

Fig. 7.29 (a) Kerr rotation data as a function of probe energy with σ^+ and σ^- polarized pump at $V_b = 0.2\,V$, in zero magnetic field. The PL at this bias is also shown (solid line), and the X^- and X^0 energies are indicated by the dotted lines. (b) The sum of the two curves shown in (a), representing spin-independent effects, such as the spike (labeled Ξ^0) at the X^0 energy. (c) The difference of the two curves shown in (a), representing the spin dependent signal. The feature Ξ^- at the X^- energy is attributed to single spin detection.

Ξ^0 at the X^0 energy clearly does not depend on the sign of the injected spin and is similar to features seen in single dot absorption measurements [23]. This peak may be due to polarization dependent absorption in the QD. We focus here on the $(\sigma^+ - \sigma^-)$ data, which represents KR due to the optically oriented spin polarization. The feature Ξ^- at the X^- energy only appears in the difference data, indicating that it is due to the injected spin polarization, shown in Figure 7.30 at four different bias voltages. For all voltages, the Ξ^- feature is centered at the X^- transition energy, indicated by the triangles. These data can be fit to Eq. (7.5) including only a single transition in the sum, on top of a broad background (solid lines, Figure 7.30). From the free parameters in these fits various parameters can be determined: the transition energy E_0, amplitude A (defined as half the difference of the local maximum and minimum near E_0), and width Γ of the Ξ^- Kerr rotation feature.

By measuring this Kerr rotation spectrum around the X^- resonance in a quantum dot, one obtains a measurement of the spin polarization of the single electron state. Hanle-type measurements can be obtained using this technique by applying a transverse magnetic field. The typical Lorentzian lineshapes are observed, though the analysis is somewhat complicated by the presence of multiple lifetimes, as in Section 7.3.2.

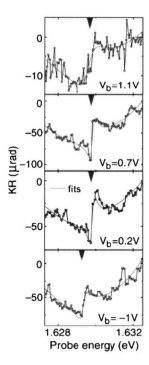

Fig. 7.30 Single spin Kerr rotation feature as a function of bias voltage. The Ξ^- KR feature is present over a large range of bias voltage, though it broadens and decreases in amplitude as the charging increases. Lines are fits to an odd-Lorentzian plus a broad (Gaussian) background. Triangles indicate the energy of the X^- transition, determined from PL measurements.

7.3.7
Time-Resolved Observation of Single Spin Coherence

The ability to sequentially initialize, manipulate, and read out the state of a qubit, such as an electron spin in a quantum dot, is a requirement in virtually any scheme for quantum information processing [15, 20, 265]. However, the optical measurements of a single electron spin described in the previous section have focused on time-averaged detection, with the spin being initialized and read out continuously. In the work of M. H. Mikkelsen *et al.*, "Optically detected coherent spin dynamics of a single electron in a quantum dot" [238], the measurement scheme of the previous section is modified to directly observe the coherent evolution of an electron spin in a single dot, using time-resolved Kerr rotation spectroscopy. This all-optical, nondestructive technique allows one to monitor the precession of the spin in a superposition of Zeeman-split sublevels with nanosecond time resolution.

In the present work, as in Section 7.3.6, the electron is confined to a single interface fluctuation quantum dot. The dot is embedded within a diode structure, allowing controllable charging of the dot with a bias voltage [40]. Also, the QD is centered within an integrated optical cavity to enhance the small, single spin KR signal [60]. See Figure 7.27 for a schematic of the sample structure. With circularly polarized excitation, spin-polarized electrons and holes are pumped into the quantum well, according to the selection rules governing interband transitions in GaAs [121]. One or more electrons and/or holes then relax into the dot. By measuring the subsequent single dot photoluminescence, the equilibrium charge state of the quantum dot as well as the energies of various interband optical transitions as a function of bias voltage can be determined [60, 211]. The measurements described below are performed at a bias voltage where the dot is nominally uncharged, and the optical excitation injects one or more electrons or holes. In this regime, the dot may contain a single spin-polarized electron through the capture of an optically injected electron, or spin dependent X^- decay [60]. Knowing the transition energy E_{X^-} from the PL measurements, the spectroscopic dependence of the Kerr effect is used to isolate the dynamics of the single electron spin from that of multiparticle complexes, such as charged or neutral excitons.

The setup used for time-resolved single spin measurements is similar to the setup for the continuous measurement, but with a few important modifications. Instead of the continuous pump laser, a mode-locked Ti:Sapphire laser provides pump pulses with energy $E_{pump} = 1.653$ eV, and duration $\sim$150 fs at a repetition period $T_r = 13.1$ ns. The bandwidth of the spectrally broad pump pulses is narrowed to $\sim$1 meV by passing the pump beam through a monochromator. The probe pulses are derived from the same wavelength tunable continuous wave Ti:Sapphire laser as in Section 7.3.6. However, the probe laser is now passed through an electro-optic modulator, allowing for electrical control of the probe pulse duration from cw down to 1.5 ns. The modulator is driven by an electrical pulse generator triggered by the pump laser, allowing for electrical control of the time delay between the pump and the probe pulses with picosecond precision. This technique yields short pulses while maintaining the narrow linewidth and wavelength tunability of the probe laser. With these pulsed pump and probe lasers, a time-resolved measurement may be performed, as described in Section 7.3.4. As in the time-resolved measurements described above, though the signal is averaged for several seconds (the spin is re-initialized and probed millions of times), the stroboscopic pump/probe technique allows measurement with high time resolution.

For a fixed delay between the pump and the probe, the Kerr rotation angle, θ_K, is measured as a function of probe energy. At each point, the pump excitation is switched between σ^+ and σ^- polarization, and the spin dependent signal is obtained from the difference in θ_K at the two helicities. The resulting Kerr rotation spectrum is fit to a single term of Eq. (7.5) plus a constant vertical offset, y_0. The amplitude, θ_0, of the odd-Lorentzian is proportional to the projection of the spin in the dot along the measurement axis. By repeating this measurement at various pump-probe delays, the evolution of the spin state can be mapped out.

The single spin Kerr rotation amplitude as a function of delay, measured with a 3-ns duration probe pulse and a magnetic field $B = 491$ G, is shown in Figure 7.31a, exhibiting the expected oscillations due to the coherent evolution described above. Figure 7.31b–f shows a sequence of Kerr rotation spectra at several delays, and the fits from which the data in Figure 7.31a are obtained. In the inset of Figure 7.31a the offset y_0 is shown, which oscillates with the same frequency as the single spin Kerr rotation but decays with a shorter lifetime. This behavior may be consistent with that of free electron spins in the quantum well, previously investigated in

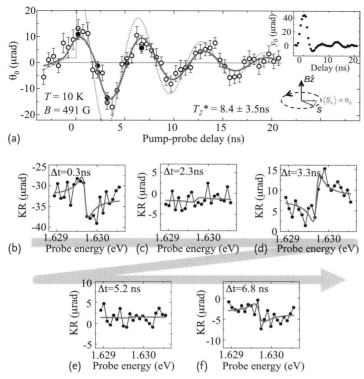

Fig. 7.31 (a) As the pump-probe delay is varied, coherent precession is seen as oscillations in the measured projection of the spin. The solid curve is a fit to the model and the dotted line shows the same curve but without the probe pulse convolution. The inset shows the offset, y_0 as a function of delay. The error bars are the standard error in the fits to the Kerr rotation spectra. (b)–(f) KR spectra at increasing pump-probe delay showing amplitude oscillations in time. The solid curves are the fits from which the solid points in part (a) were obtained.

time-averaged measurements [258]. Due to the small confinement energy of these dots (several meV) relative to the quantum well, one does not expect a significant shift in the g-factor between the quantum dots and the well.

The solid line in Figure 7.31a is a fit to a simple model of spin precession (see [238]), from which values of the precession frequency and effective coherence time T_2^* may be extracted. The model includes a convolution with the measured profile of the 3-ns-duration probe pulse, which smears out the features to some extent. The dashed line shows the model curve without the probe pulse convolution, plotted with the same parameters for comparison.

In Figure 7.32a the precession of the spin is shown at three different magnetic fields. As expected, the precession frequency increases with increasing field. The solid lines in Figure 7.32a are fits to the model, and the frequency Ω obtained from such fits is shown in Figure 7.32b as a function of magnetic field. A linear fit to these data yields an electron g-factor of $|g| = 0.17 \pm 0.02$, consistent with

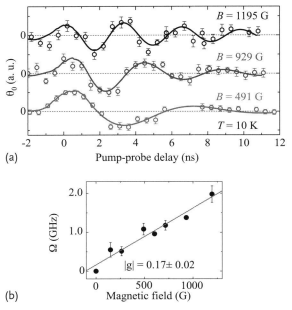

Fig. 7.32 (a) Spin precession at three magnetic fields. The error bars are the standard error in the fits to the Kerr rotation spectra. (b) Precession frequency, Ω, as a function of magnetic field. The error bars represent the standard deviation from repeated measurements. The linear fit yields a g-factor of $\pm 0.17 \pm 0.02$.

the range of g-factors for these quantum dots found in previous ensemble or time-averaged measurements [211, 266]. At zero magnetic field, as shown in Figure 7.33, the spin lifetime is found to be $T_2^* = 10.9 \pm 0.5$ ns. This value agrees with previous time-averaged [211] and ensemble [258, 266] measurements where the relevant decay mechanism is often suggested to be dephasing due to slow fluctuations in the nuclear spin polarization. However, these polarization fluctuations are not expected to result in a single exponential decay of the electron spin [226, 227], suggesting that decay mechanisms other than nuclear spin fluctuations might also be relevant in this case. In these quantum dots, the electronic level spacing of ~ 1 meV [24] is of the same order as $k_B T$ for this temperature range. Therefore, thermally-activated or phonon-mediated processes [190, 249, 267, 268] which yield an exponential decay, might be significant in this regime.

These measurements constitute a noninvasive optical probe of the coherent evolution of a single electron spin state with nanosecond temporal resolution, which is a key ingredient for many spin/photon-based quantum information proposals [269, 270]. Furthermore, this technique provides a sensitive probe of the dynamics of the spin, revealing information about the spin coherence time and g-factor. Future work may exploit this ability to further explore the relevant decoherence mechanisms and the electron-nuclear spin interactions.

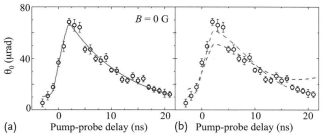

Fig. 7.33 (a) Decay of the spin polarization with $B = 0$. The solid line is a single exponential fit convolved with the probe pulse, giving a reduced $\chi^2 = 3.8$. The error bars are the standard error from the fits to the KR spectra. (b) The same data as (a) with a fit to a model of nuclear spin dephasing, yielding an obviously poorer fit, and a reduced $\chi^2 = 24.3$. A fit to the same model multiplied by an additional exponential decay factor, to model both nuclear dephasing and other decoherence mechanisms, is also shown. This fit is also significantly worse than that in (a).

7.3.8
Optical Spin Manipulation

Using ultrafast optical pulses to coherently manipulate the spin state of an electron is a key ingredient in many proposals for solid state quantum information processing [180, 181, 271–274]. Though electrical control of single spins has been achieved [260, 261], the nanosecond timescales required for such manipulation limits the number of operations that can be performed within the spin coherence time. For example, single electrons were confined to a gate-defined 2DEG dot, and the spin control was achieved via spin resonance induced by a stripline deposited on the sample. The speed of such a spin rotation is limited by the maximum attainable AC magnetic field. (Electron spin resonance is also possible in optically active quantum dots. This was described theoretically [188], and then observed experimentally [261]. However, the optical techniques used in such experiments have been described above, and the ESR is essentially an electrical technique so will not be discussed further here.) On the other hand, spin control via picosecond-scale optical pulses yields an improvement of several orders of magnitude in the manipulation time. In J. Berezovsky *et al.* "Picosecond coherent optical manipulation of a single electron spin in a quantum dot" [239], such a scheme for a single electron spin in a quantum dot is demonstrated, monitoring the coherent evolution of the spin state using time-resolved Kerr rotation spectroscopy. The spin is subjected to an intense, off-resonant laser pulse, which induces a rotation of the spin through angles up to π radians on picosecond timescales.

The optical (or ac) Stark effect (OSE) was first studied in atomic systems in the 1970s [275–277] and subsequently explored in bulk semiconductors and in quantum wells [278–280]). In recent years, OSE has been used to observe ensemble spin manipulation in a quantum well [281] and to control orbital coherence in a quantum dot [282]. Additionally, other optical manipulation schemes have been explored on ensembles of spins [283, 284]. As described in Section 5.6, it is found that an

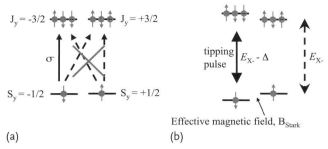

(a)   (b)

Fig. 7.34 (a) The quantum dot transitions relevant to the optical Stark effect, illustrated in the basis along the *y* axis (the growth direction). The single electron ground state is coupled optically to the negatively charged exciton (trion) state. For a given circular polarization, the selection rules allow a nonzero matrix element only for one such transition. (b) When an off-resonant circularly polarized optical field is applied, one spin state is shifted due to the OSE. This results in a spin splitting (effective magnetic field) for the single electron.

optical field with intensity I_{tip}, detuned from an electronic transition by an energy Δ, induces a shift in the transition energy

$$\Delta E \approx \frac{d^2 \, I_{\text{tip}}}{\Delta \, \sqrt{\epsilon/\mu}} \, , \tag{7.6}$$

where d is the dipole moment of the transition, and ϵ and μ are the permittivity and permeability of the material [279]. Figure 7.34 shows the relevant energy levels for the quantum dot system considered here. The ground state consists of a single electron in the lowest conduction band level, spin-split by a small magnetic field, B_z. The lowest energy interband transition is to the trion state consisting of two electrons in a singlet state and a heavy hole. Due to the optical selection rules (see Chapter 5), the dipole strength of this transition in the basis along the *y* direction from the spin-up (-down) ground state is zero for σ^+ (σ^-) polarized light, as indicated in the diagram. Therefore, for circularly polarized light, the OSE shifts just one of the spin sublevels and produces a spin splitting in the ground state, which can be represented as an effective magnetic field, B_{Stark}, along the light propagation direction. By using ultrafast laser pulses with high instantaneous intensity to provide the Stark shift, large splittings can be obtained to perform coherent manipulation of the spin within the duration of the optical pulse (here, $B_{\text{Stark}} \sim 10\,\text{T}$). Note that this phenomenon can also be described in terms of a stimulated Raman transition [180, 272], or as an avoided crossing between excitons and photons [285].

As in the previous two sections, the sample consists of a layer of charge-tunable GaAs interface QDs embedded in an optical cavity (see Figure 7.27). A schematic of the experimental setup is shown in Figure 7.35, in all its gory detail. This is similar to the setup of the last two chapters, but again, with some significant changes. In this case, three synchronized, independently tunable optical pulse trains are focused onto the sample: the pump, the probe, and the "tipping pulse" (TP). The

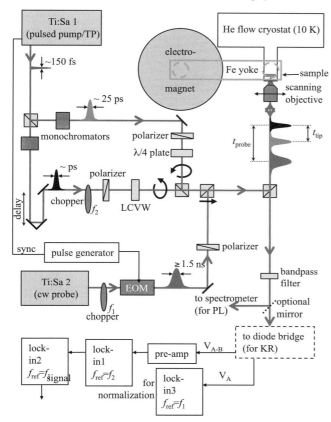

Fig. 7.35 Setup for Stark tipping measurements.

pump and tipping pulse are both derived (by means of a beamsplitter) from the mode-locked Ti:Sapphire laser generating a train of $\sim$ 150-fs-duration pulses at a repetition rate of 76 MHz. The pump is circularly polarized and tuned to an energy $E = 1.646$ eV (FWHM $\sim$1 meV), thereby injecting spin-polarized electrons and holes into the continuum of states above the dot [121]. One or more of these electrons or holes can then relax into the dot. The circularly polarized TP (duration $\sim$30 ps, FWHM $= 0.2$ meV) is tuned to an energy below the lowest quantum dot transition and is used to induce the Stark shift. The relative time delay between the pump pulse and the TP is controlled by a mechanical delay line in the pump path.

As in the previous section, the probe pulse is generated by passing a narrow linewidth continuous-wave laser through an electro-optic modulator synchronized with the pump/TP laser. The resulting 1.5-ns-duration pulses probe the spin in the QD through the magneto-optical Kerr effect [238].

In a typical measurement, the pump pulse arrives at $t = 0$ along the y axis (growth direction), and in some cases, a single spin-polarized electron will relax into the quantum dot. For pump helicity $\sigma^{\pm}$, this electron is (up to a global phase) initially in the state

$$|\psi(t=0)\rangle = \frac{1}{\sqrt{2}}(|\uparrow\rangle \pm i|\downarrow\rangle)\,, \tag{7.7}$$

where "up" and "down" are chosen as the basis states along the external magnetic field B_z. The spin then begins to coherently precess at the Larmor frequency $\omega = g\mu_B B_z/\hbar$:

$$|\psi(t)\rangle = \frac{1}{\sqrt{2}}\left(e^{-i\omega t/2}|\uparrow\rangle \pm i e^{i\omega t/2}|\downarrow\rangle\right)\,, \tag{7.8}$$

where g is the effective electron g-factor, μ_B is the Bohr magneton, and $\hbar$ is the reduced Planck constant. At time $t = t_{\text{tip}}$, the TP arrives and generates an additional spin splitting along the y axis for the duration of the pulse. During this time, the spin precesses about the total effective field (which is typically dominated by B_{Stark}), and then continues to precess about the static applied field. The probe then measures the resulting projection of the spin in the dot, $\langle S_y \rangle$ at $t = t_{\text{probe}}$. This sequence is repeated at the repetition frequency of the laser (76 MHz), and the signal is averaged for several seconds for noise reduction. As described in Section 7.3.4, the pump and probe are modulated using mechanical choppers, allowing for lock-in detection to measure only spins that are injected by the pump. Also, the pump is switched between σ^+ and σ^-, with a measurement made at each helicity. The spin signal is then taken as the difference between these values, eliminating any spurious signal from spins not generated by the pump (for instance, phonon-assisted absorption from the TP [286]), or non-spin-dependent rotation of the probe polarization.

It is convenient to understand the observed spin dynamics in the Bloch sphere picture, described in Chapter 1. Here, the spin state is represented as a vector (S_x, S_y, S_z), where $(0, 0, \pm S_z)$ represents the eigenstates $|\uparrow\rangle$ and $|\downarrow\rangle$, and vectors with nonzero S_x and S_y represent coherent superpositions of $|\uparrow\rangle$ and $|\downarrow\rangle$. In this picture, the dynamics of the spin can be calculated by applying the appropriate sequence of rotation matrices to the initial state. Figure 7.36 illustrates the sequence of rotations described by the model.

The initial spin state at $t = 0$ is taken to be

$$\mathbf{S}_0 = \begin{pmatrix} 0 \\ S_{0,y} \\ S_{0,z} \end{pmatrix}\,, \tag{7.9}$$

where the initial component in the z direction, $S_{0,z}$, is assumed to be small due to misalignment of the pump beam from normal incidence. Before the tipping pulse arrives, the spin freely precesses around the applied field. At $t = t_{\text{tip}}$, the tipping pulse rotates the spin through an angle ϕ_{tip} about the y axis. Since the duration of the TP is much less than ω^{-1}, the tipping is assumed to occur instantaneously. At $t > t_{\text{tip}}$, this state $\mathbf{S}(t_{\text{tip}})$ then precesses freely about the z axis at the Larmor frequency.

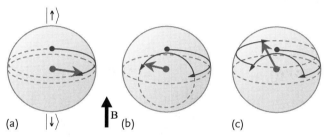

Fig. 7.36 Sequence of rotations in the Stark tipping model.
(a) Before TP the spin is approximately along the y axis and
then precesses about the magnetic field, B_z. (b) At $t = t_{\text{tip}}$,
the spin is instantaneously rotated about the y axis through an
angle ϕ_{tip}. (c) After TP, the spin continues to precess about the
magnetic field.

This model can be extended further to include the effects of spin decoherence,
as well as dynamic nuclear polarization (see [239]).

Figure 7.37a shows the time evolution of a single spin in a transverse magnetic
field, with no TP applied. Each data point is determined from the fit to a KR spectrum at a given pump-probe delay. Using the model outlined above, a least-squares
fit to this data can be performed to determine various parameters in the model: ω,
T_2^*, and the effective field from the nuclear polarization, B_n. The gray curve in Figure 7.37a shows the result of this fit, and the dotted line is the corresponding plot
of the model without probe pulse convolution. As expected, the spin is initialized
at $t = 0$, and then precesses freely about the applied field.

The data in Figure 7.37b and c show the same coherent spin dynamics of Figure 7.37a, but with the TP applied at $t = t_{\text{tip}}$. The intensity of the TP is chosen to induce a 1.05π rotation about the y axis. In Figure 7.37b, the TP arrives
at $t_{\text{tip}} = 1.3\,\text{ns}$, when the projection of the spin is mainly along the x axis. This
component of the spin is thus rotated by the TP through $\sim \pi$ radians. The predicted spin dynamics as given by the model is shown in the dotted line, and the same
curve convolved with the probe pulse is given by the solid line. Note that this curve
is not a fit – all of the parameters are determined either in the fit to Figure 7.37a,
or elsewhere. Only the overall amplitude of the curve has been normalized. Here,
the spin is initialized at $t = 0$, and as before, precesses freely until the arrival of
TP. After TP, the spin has been flipped and the resulting coherent dynamics show
a reversal in sign. This can be clearly seen by comparing the sign of the measured
signal at the position indicated by the dashed line in Figure 7.37.

Figure 7.37c shows the spin dynamics again with the same parameters, but with
$t_{\text{tip}} = 2.6\,\text{ns}$. The spin at this delay will have only a small projection in the x–z
plane and the TP-induced rotation about the y axis will have only a small effect on
the spin state. This expectation is borne out in the data, where the spin dynamics
show essentially the same behavior as in the absence of the TP (Figure 7.37a).
Again, the model yields qualitatively the same behavior.

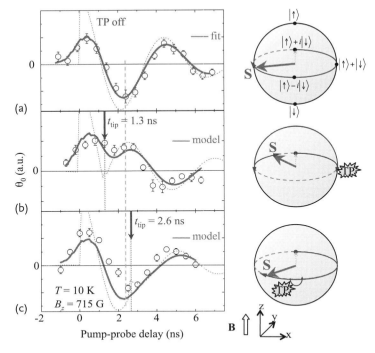

Fig. 7.37 (a) Coherent spin precession with no TP. Error bars indicate the standard error in the fits to the KR spectra. The solid curve is a fit to the model convolved with the probe pulse; the dotted line is the same, without the probe pulse convolution. (b) and (c) Same conditions as (a) but with TP applied at $t_{tip} = 1.3$ ns and $t_{tip} = 2.6$ ns, respectively, with intensity $I_{tip} = 4.7 \times 10^5$ W/cm^2 and detuning $\Delta = 2.65$ meV, to induce a 1.05π rotation. The solid curves in (b) and (c) are from the model, using parameters obtained elsewhere. The gray dashed line highlights the change in sign of the spin precession in (b) as compared to (a) and (c). The diagrams on the right illustrate the effect of TP on the spin dynamics.

These results demonstrate the ability to coherently rotate a single electron spin through angles up to π radians on picosecond timescales. A simple model including interactions with nuclear spins reproduces the observed electron spin dynamics with a single set of parameters for all of the measurements. In principle, at most 200 single qubit flips could be performed within the measured T_2^* of 6 ns. However, by using shorter tipping pulses and QDs with longer spin coherence times, this technique could be extended to perform many more operations within the coherence time. A mode-locked laser producing ~ 100-fs-duration tipping pulses could potentially exceed the threshold ($\sim 10^4$ operations) needed for proposed quantum error correction schemes [265]. Additionally, this spin manipulation technique may be used to obtain a spin echo [287], possibly extending the observed spin coherence time, as in Section 7.3.5.

7.3.9
Putting It All Together

Though real quantum information processing has yet to be demonstrated in an optical quantum dot spin-based scheme, the recent work of D. Press *et al.*, "Complete quantum control of a single quantum dot spin using ultrafast optical pulses" [240] demonstrates the current state of the art in putting together the various necessary elements. In this work, high fidelity optical spin pumping, ultrafast optical manipulation, and optical spin readout are all combined in one experiment.

First, spins are pumped into a singly-charged InAs self-assembled quantum dot in essentially the same way as described above in Atatüre *et al.* [161], except here the magnetic field is applied in the plane of the sample, and thus the excitation light is polarized linearly instead of circularly. This serves to pump spins into a Zeeman-split eigenstate, in a direction perpendicular to the laser direction. Here, the pump laser is on continuously throughout the experiment, but only has a small effect during the manipulation and measurement of the spin. This optical pumping yields an initialization fidelity of about 92%.

The spin of the electron in the dot is measured using a photoluminescence technique analogous to the absorption technique used in the Atatüre work. Instead of measuring the absorption of the pumping transition, here the spin is measured by the luminescence emitted while the pumping transition is being driven. Once the spin has been pumped into the initialized state, this transition is no longer driven and no luminescence is detected. If the spin is rotated back into the other state (or has some projection onto the other state) then the pumping laser excites its transition again and photons are detected. Thus no luminescence signal means the spin is in the initialized state, and maximal luminescence means the spin is in the opposite state.

Once the spin is initialized, it may be then manipulated using an off-resonant tipping pulse, as described in the previous section. Here, the tipping pulse is about

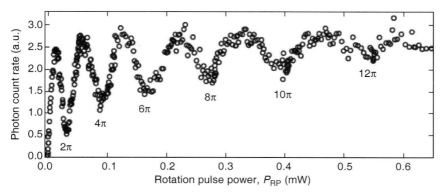

Fig. 7.38 Coherent manipulation of a single spin showing rotations up to 13π. Reprinted by permission from Macmillan Publishers Ltd.: Nature [240], copyright (2008).

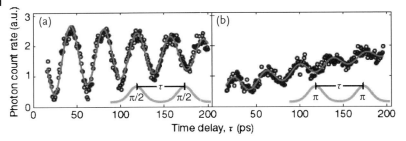

Fig. 7.39 Ramsey fringes of a single electron spin in a quantum dot. (a) A $\pi/2$ pulses rotate the spin to be perpendicular to the magnetic field, which then precesses for a time τ before a second $\pi/2$ pulse is applied. (b) A pair of π pulses are applied separated by time τ. Only small fringes are observed. Reprinted by permission from Macmillan Publishers Ltd.: Nature [240], copyright (2008).

4 ps in duration (that is, the manipulation of the spin takes place in $\sim$4 ps). As above, by increasing the tipping pulse intensity, the spin may be rotated through increasingly large angles. Figure 7.38 shows the spin, as monitored by the luminescence, being rotated by the tipping pulse as a function of tipping pulse intensity. The first π rotation is found to have fidelity of 91%.

This experiment is further extended by adding a second tipping pulse applied a controllable time, τ, after the first. This allows the observation of so-called Ramsey fringes. In Figure 7.39a the spin is measured after the application of a pair of $\pi/2$ ultrafast rotations, separated by a time τ. The spin, initialized into an eigenstate of the magnetic field, is rotated by the first pulse to be perpendicular to the magnetic field. The spin then precesses freely until the arrival of the next pulse. If the spin has precessed a full whole number of times, the second $\pi/2$ pulse will rotate the spin the rest of the way to be opposite to the original direction. On the other hand, if the spin has precessed a whole number of times plus a half precession, the spin will be rotated back to its initial direction. The result is that, as the time between the pulses is swept, the signal oscillates between the maximum and minimum, with different coherent states being generated in between.

Figure 7.39b shows the same thing as part 7.39a, only with a pair of π pulses. Here, the first π pulse rotates the spin into the opposite eigenstate, which then ideally remains there until the next pulse rotates the spin back to the initial state. This results in no observed fringes – the small fringes are present only due to small nonidealities in the experiment.

By adjusting the timing and intensities of these two tipping pulses, the spin can be rotated into any arbitrary state. This complete access to any point on the Bloch sphere, with high fidelity and picosecond operation times, coupled with high initialization fidelity represents real progress towards optically-addressed, spin-based quantum information processing in quantum dot systems. One of the main challenges that remains is the scaling up of these systems to many coupled qubits. One possible solution is proposed by S. M. Clark *et al.* [180], by coupling a number

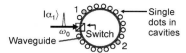

Fig. 7.40 Proposed scheme for scalable quantum computing with optically controlled spins in quantum dots. Single qubit rotations are performed using the methods described in Sections 7.3.8 and 7.3.9 while two-qubit operations are performed via light in the waveguide through an effect analogous to the Faraday effect discussed in Section 7.3.4. Reprinted with permission from [180]. Copyright (2007) by the American Physical Society.

of quantum dots together via an integrated waveguide. This proposed scheme is shown schematically in Figure 7.40. The quantum dots are shown arranged around a circular waveguide, with each dot sitting within an optical cavity (potentially an integrated photonic crystal cavity, see Section 4.2) to enhance the interaction between light and the dots. Single qubit operations may be performed as described above in the previous two sections. Two-qubit operations (or more) might be effected by injecting light into the waveguide that could then couple any resonant quantum dots via a conditional phase shift, related to the Faraday effect described in Section 7.3.4.

Whether this proposal turns out to be possible or not, the methods and results described in this chapter provide an indication that there is significant potential for quantum information science and possible applications involving spins in optically active quantum dots.

8
Controlling Charge and Spin Excitations in Coupled Quantum Dots

In the "artificial atom" picture of quantum dots, a set of coupled quantum dots reminds us of an artificial molecule. Each quantum dot offers orbitals that are well localized in space, and the tunneling of electrons as well as the electromagnetic interaction provide coupling mechanisms just as in the smaller molecular counterparts.

In this chapter we consider a pair of coupled quantum dots, which we refer to as a quantum dot molecule (QDM). We start this chapter with a simple model for a single electron that is delocalized in a quantum dot molecule. The delocalized single-electron states are then shown to be observable in the optical spectrum of a single exciton, X^0. We then consider two-electron states in the quantum dot molecule and review established theoretical models from molecular physics, which are then applied to interpret recent experimental work.

8.1
Tunable Coupling in a Quantum Dot Molecule

The model of a particle hopping between two orbitals $|\psi_i^0\rangle$, where $i = 1, 2$, is well known from quantum mechanics textbooks. We start our discussion of a pair of interacting quantum dots by introducing this underlying basic concept. We then show that carriers in a quantum dot molecule exhibit this behavior.

8.1.1
A Toy Model for Coupled Systems: The Two-Site Hubbard Model

Let us assume that the Hamiltonian in the absence of any interdot coupling is H^0 and the orbitals have eigenenergies E_i^0, such that

$$H^0|\psi_i^0\rangle = E_i^0|\psi_i^0\rangle .$$

(8.1)

Spins in Optically Active Quantum Dots. Concepts and Methods.
Oliver Gywat, Hubert J. Krenner, and Jesse Berezovsky
Copyright © 2010 WILEY-VCH Verlag GmbH & Co. KGaA, Weinheim
ISBN: 978-3-527-40806-1

After introducing a transfer term t, which couples the two dots, the new eigenstates can be written as a linear combination of the uncoupled states

$$H|\psi_i\rangle = E_i|\psi_i\rangle\ , \tag{8.2}$$

$$|\psi_i\rangle = \alpha|\psi_1^0\rangle + \beta|\psi_2^0\rangle\ , \tag{8.3}$$

where the numerical parameters α and β are calculated in the following. The resulting coupled Hamiltonian can be written in matrix notation as

$$H \doteq \begin{pmatrix} E_1^0 & t_{12} \\ t_{12}^* & E_2^0 \end{pmatrix}\ , \tag{8.4}$$

with the coupling included by the transfer matrix elements $\langle\psi_2^0|t|\psi_1^0\rangle = t_{21} = t_{12}^*$.

After diagonalization, the eigenvalues of the coupled system are obtained,

$$E_\pm = E_1^0 + \frac{1}{2}\Delta \pm \frac{1}{2}\sqrt{\Delta^2 + |2t_{12}|^2}\ , \tag{8.5}$$

where the energy difference between the uncoupled eigenstates is $\Delta = E_2^0 - E_1^0$. The coefficients α and β in Eq. (8.3) satisfy the normalization condition $|\alpha|^2 = 1 - |\beta|^2$, and, explicitly,

$$|\alpha|^2 = \frac{|2t_{12}|^2}{|2t_{12}|^2 + (\Delta \pm \sqrt{\Delta^2 + |2t_{12}|^2})^2}\ . \tag{8.6}$$

The coupled eigenenergies $E_\pm$ as a function of Δ are shown schematically in Figure 8.1. Clearly, for a degenerate system ($\Delta = 0$) we obtain the minimum energy splitting,

$$\min|E_- - E_+| = |2t_{12}|\ , \tag{8.7}$$

which is determined by the coupling strength alone. The corresponding eigenstates are the symmetric and antisymmetric superpositions of the uncoupled states,

$$|\psi_-\rangle = \frac{1}{\sqrt{2}}\left(|\psi_1^0\rangle + |\psi_2^0\rangle\right)\ , \tag{8.8}$$

$$|\psi_+\rangle = \frac{1}{\sqrt{2}}\left(|\psi_1^0\rangle - |\psi_2^0\rangle\right)\ , \tag{8.9}$$

referred to as the bonding ($|\psi_-\rangle$) and the antibonding ($|\psi_+\rangle$) molecular states. Here, the index of the wave function refers to the associated energy $E_\pm$. As the degeneracy is lifted ($\Delta \neq 0$), the eigenstates evolve from completely hybridized states asymptotically into the uncoupled states $|\psi_i^0\rangle$ with increasing detuning. This behavior is characteristic for tunable coupled quantum states and is called avoided crossing or anticrossing. Therefore, its observation is a direct proof and characteristic fingerprint of coherent coupling between two quantum states.

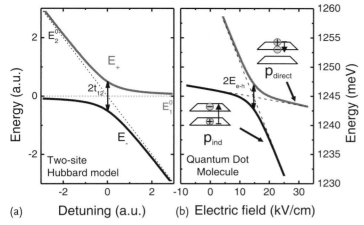

(a) **Detuning (a.u.)**   (b) **Electric field (kV/cm)**

Fig. 8.1 Level anticrossings as a signature of coherent coupling. (a) Two-site Hubbard model with eigenenergies $E_{\pm}$ provided by Eq. (8.5) as a function of the energy level detuning $\Delta = E_2 - E_1$. With increasing absolute value of the detuning, the energies evolve asymptotically into the uncoupled eigenvalues E_2^0 and $E_1^0 = 0$. The minimum splitting is given by $|2t_{12}|$ for $\Delta = 0$ for the completely hybridized states. (b) Realistic calculation for the lowest energies of an exciton in a QDM as a function of an applied electric field, taking into account electron tunneling and the Coulomb interaction. The asymptotic states are here the direct and indirect exciton state, respectively.

We underline that at this point no assumption has yet been made about the nature of the coupling that is captured by the transfer term t. Several types of coupling are possible for quantum dots. Ouyang and Awschalom have observed coherent spin transfer between quantum dots coupled by a benzene ring [288]. The experimental results have been successfully modeled with a similar model to that outlined above by Meier *et al.* [289], assuming a direct carrier transfer via the coupling molecule. These results were in reasonable agreement with atomistic calculations by Schrier and Whaley [290].

As another example, resonant Förster transfer is a process that can transfer an exciton from one quantum dot to another [291, 292]. Without going into further detail here, Förster transfer results from the electrostatic dipole interaction Hamiltonian

$$t_F = \frac{e^2}{4\pi\epsilon R^3}\left[\langle \mathbf{r}_1\rangle\langle \mathbf{r}_2\rangle - \frac{3}{R^2}\left(\langle \mathbf{r}_1\rangle \cdot \mathbf{R}\right)\left(\langle \mathbf{r}_2\rangle \cdot \mathbf{R}\right)\right], \tag{8.10}$$

where $\mathbf{R}$ denotes the relative position of the quantum dots and $\langle \mathbf{r}_{1,2}\rangle$ are the dipole transition matrix elements between the exciton vacuum and the one-exciton states in dots 1 and 2, respectively. Govorov has provided detailed calculations for the Förster transfer between quantum dots of various shapes [291]. Nazir *et al.* have studied in further detail the Förster coupling of three-dimensional parabolic quantum dots [292]. The matrix elements for Förster transfer between nested spherical quantum shells have been calculated in [205]. In general, if the symmetry axes of both quantum dots are aligned, then the Förster transfer in Eq. (8.10) conserves the exciton spin [291, 292].

As discussed in Section 5.6, the dispersive interaction with a cavity mode can provide another mechanism to couple exciton or spin states of spatially separated quantum dots, with a maximum strength $t_{\text{Disp}} = \Omega^2(1/\Delta_1 + 1/\Delta_2)$ for excitons, where Δ_i refers to the detuning of the cavity mode with respect to the exciton transition of dot i, and Ω is the optical Rabi frequency.

Direct tunneling is a fourth possibility for quantum dot coupling. In this case, orbitals of the two dots have finite overlap, which allows for quantum tunneling between the two sites. Burkard and coworkers have provided molecular theory calculations for two electrons in laterally [138] and vertically [140] coupled quantum dots, as well as for a biexciton in laterally coupled quantum dots [139]. Berezovsky and coworkers have investigated the coupling of a core and a shell in a spherical heterostructure [132]. In the following we highlight experiments that have shown that the states of coupled quantum dots can be understood in detail with theories adapted from molecular physics.

8.1.2
An Exciton in a QD Molecule: A Coupled System

In coupled QDs studied in optical experiments, interband excitations, that is, excitons are probed. We are now going to describe these basic charge excitations and their manipulation using static electric fields. In order to put the simple model shown above into practice we have to determine the detuning Δ between QD levels and its underlying tuning mechanism.

Let us consider a system consisting of two ($i = 1, 2$) vertically stacked quantum dots with height $h_{\text{QD},i}$ which are separated by a barrier of thickness d_{id}, to which we refer to as the interdot distance. When subject to an axial electric field (parallel to the QD stack), the energies of the confined uncoupled states of the two QDs are tuned relative to each other due to the Stark effect. The resulting energy level difference is

$$\Delta = \left(d_{id} + \frac{h_{\text{QD},1}}{2} + \frac{h_{\text{QD},2}}{2} \right) F = \delta \cdot F \, , \tag{8.11}$$

where we call δ the interdot electrostatic lever arm. This equation allows for conversion of the electric field F into the detuning Δ. Using Eqs. (8.6) and (8.11), the energy splitting of the coupled QD system as a function of F can be calculated as

$$\Delta E = \sqrt{(2E_m)^2 + \delta^2(F - F_m)^2} \, , \tag{8.12}$$

where $2E_m$ is the coupling energy for a pair of states labeled m in the two QDs, which are brought into resonance at their corresponding "critical" resonance field F_m.

In an interband optical experiment such as PL spectroscopy, the radiative emission of excitons is observed. Let us consider the fundamental interband excitation in a QDM, the charge neutral single exciton $X^0 = 1e + 1h$. It turns out that signatures of coherent coupling between the QDs can be observed in exciton PL spectra.

In general, one can distinguish between two different species of excitons in a QDM: spatially *direct* excitons, for which e and h are localized in the same dot, and spatially *indirect* excitons, for which e and h are localized in different dots. In the following, we focus on quantum coupling mediated by tunneling of the electron part of the exciton wavefunction. Even though coupling between hole levels has raised a lot of interest recently [293–296], it is typically much weaker than for electron levels due to the smaller spatial extension of hole wave functions. The electron levels of the two dots are detuned relative to each other by changing F. At the corresponding resonance field F_{1e+1h} these levels are degenerate. Now, taking into account the tunnel coupling through the thin barrier between the two QDs, we obtain hybridization of the electron part of the exciton wavefunction into bonding and antibonding orbitals, as outlined in Section 8.1.1. The resulting bonding and antibonding exciton branches are called X_B and X_A in the following, respectively.

This model at present only accounts for the coupled e levels in the QDM. We now include the total electrostatic interaction with F and the Coulomb interaction between e and h. As discussed in Section 4.1.4, an exciton localized in a nanostructure subject to a static electric field F exhibits a shift due to the quantum confined Stark effect (QCSE). This shift is given by $-pF$, where $p = ed_{eh}$ is the electrostatic dipole moment of the exciton, with the separation d_{eh} of the e and h wavefunction centers within the same dot. In a single QD, the e–h separation is limited by the height of the dot (which is typically around $d_{eh} = 0.5$ nm). In contrast, for indirect excitons the dipole moment is determined mainly by the interdot separation, d_{id}, which can be significantly larger than d_{eh}. In the following we show data for

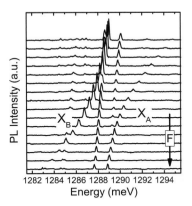

Fig. 8.2 Electric-field controlled anticrossing between spatially direct and indirect neutral excitons in a single QDM. The electric field F increases from the top to the bottom spectrum. At the resonance, the electron part of the exciton wave function hybridizes into bonding and antibonding states, giving rise to the names of the corresponding exciton branches X_B and X_A, respectively.

an example QDM consisting of two self-assembled dots separated by $d_{id} = 10\,\text{nm}$ (typical values for this type of QDs range between 4 nm and 18 nm). This gives rise to largely different dipole moments of indirect and direct excitons in the structure under study,

$$p_{\text{indir}} \simeq e \cdot 10\,\text{nm} \gg p_{\text{dir}} \simeq e \cdot 0.5\,\text{nm} . \tag{8.13}$$

Hence, the QCSE provides a straightforward way to detune the energy of direct and indirect excitons relative to each other.

In a symmetric QDM, the resonance between the dot levels would be expected at $F = 0$. However, the Coulomb interaction between e and h also has to be taken into account. Obviously, the direct exciton has a much larger attractive Coulomb matrix element than the indirect exciton. It is found that the total energy difference of the direct and the indirect exciton is actually given by $\sim$20–30 meV [70, 77]. This leads to a shift of the resonance field compared to the simple model that only considers the single-particle levels. Further shifts can be engineered by tailoring the sizes of the two dots during fabrication [71].

In Figure 8.1a we show the energies of the bonding and antibonding states in the frame of a two-site Hubbard model, described by Eq. (8.5). For comparison, Figure 8.1b shows a realistic calculation [70, 77] of spatially direct and indirect exciton energies in a pair of vertically stacked and coupled InGaAs/GaAs dots separated by 10 nm as a function of an axial electric field. In both of these cases an anticrossing is clearly visible, being an unambiguous fingerprint and direct proof of quantum coherent coupling.

A typical series of PL spectra as a function of the static electric field recorded from an individual QDM is shown in Figure 8.2. This QDM consisted of two approximately identical dots, which were embedded in an n-type Schottky diode. Taking into account the device polarity and geometry, F increases from $\sim$12 kV/cm (top spectrum) to $\sim$19 kV/cm (bottom spectrum). In the spectra, two lines labeled X_B and X_A exhibit a clear anticrossing behavior, as anticipated from the above considerations. At $\sim$16 kV/cm the resonance between the rapidly shifting indirect and the weakly shifting direct excitons is obtained. As shown in Eq. (8.7), the minimal energy splitting between the bonding and antibonding states provides the coupling energy. The anticrossing becomes even clearer when plotting the extracted peak positions as a function of F, as shown in Figure 8.3a.

For further analysis we plot the energy splitting between X_B and X_A as symbols in Figure 8.3b. At the critical electric field, $F_{\text{crit}} = F_{1e+1h} = 15.8\,\text{kV/cm}$, the minimum splitting is observed and provides a coupling energy of $E_{1e+1h} = 3.2\,\text{meV}$. Furthermore, these values are reproduced by a fit of Eq. (8.12) which is shown as a solid line in Figure 8.3b. The excellent agreement of the experimental data and the fit, which uses the adapted expression for a set of two coupled quantum states, underlines that semiconductor QDMs indeed show the behavior of an artificial molecule. In the following we elaborate on this property and introduce models from molecular physics. We also highlight how the resulting effects can be observed in a QDM.

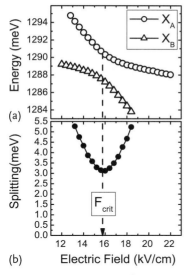

Fig. 8.3 Extracted emission line energies (a) and splitting (b) of the anticrossing in Figure 8.2.

8.2
Molecular Theory of Confined States in Coupled Quantum Dots

We have seen above that the behavior of coupled quantum dots mirrors the physics of a diatomic molecule. In the following, we consider the Heitler–London method and the Hund–Mulliken approach from molecular physics, which have proven useful for the interpretation of recent experimental work.

The Heitler–London method provides an ansatz for the molecular bond between neutral atoms. In their original work [297], Walter Heitler and Fritz London investigated the coupling of two hydrogen atoms with the nuclei fixed in space. We consider two closely spaced atoms a and b with ground state orbitals ϕ_a and ϕ_b, respectively, that have a spatial overlap integral S. The ansatz for the two-electron wave function consists of the orbitally symmetric and antisymmetric states,

$$\langle \mathbf{r}_1, \mathbf{r}_2 | \Psi_\pm \rangle = \frac{\phi_a(\mathbf{r}_1)\phi_b(\mathbf{r}_2) \pm \phi_b(\mathbf{r}_1)\phi_a(\mathbf{r}_2)}{\sqrt{2 \pm 2|S|^2}} \ . \tag{8.14}$$

Obviously, the symmetric (or bonding) wave function is a spin singlet, while the antisymmetric (or antibonding) wave function is a spin triplet. In the Heitler–London ansatz, a large on-site Coulomb repulsion is assumed for the electrons such that the double occupancy of an atom can be excluded due to energetic reasons. Here, we use the terms bonding and antibonding also for two-electron states, in addition to the delocalized single-electron states. The Heitler–London approach can be extended by including into the Hilbert space the two states in which both electrons are on the same atom,

$$\langle \mathbf{r}_1, \mathbf{r}_2 | \Psi_a \rangle = \phi_a(\mathbf{r}_1)\phi_a(\mathbf{r}_2) \quad \text{and} \quad \langle \mathbf{r}_1, \mathbf{r}_2 | \Psi_b \rangle = \phi_b(\mathbf{r}_1)\phi_b(\mathbf{r}_2) \ . \tag{8.15}$$

This approach is called the Hund–Mulliken approach (or approximation) after Friedrich Hund and Robert S. Mulliken. Obviously, for interacting electrons, the occupancy of the same dot leads to an additional energy term, the on-site Coulomb repulsion.

When adapting these models to quantum dot systems we replace the atom orbitals by the confined quantum dot states in the envelope function approximation, as introduced in Chapter 3.

8.3
Optically Probing Spin and Charge Excitations in a Tunable Quantum Dot Molecule

We return to our study of PL spectra of QDMs. At this point we take into account molecular theory for the states containing two electrons, namely, the two-electron state and the negatively charged exciton X^{1-}. These states are the initial or final states of the radiative decays shown at the center and the right of Figure 8.4. Energy splittings due to two-electron interactions in the QDM leave a characteristic footprint in the respective transition energies.

8.3.1
Optical Response with Initial and Final State Couplings

We have shown in Section 8.1 that coherent coupling in a QDM gives rise to anti-crossings of two spectral lines that belong to the coupled states. The anticrossing of the bonding and antibonding exciton states X_A and X_B discussed in Section 8.1.2 arises from a coupling in the *initial* state of the radiative decay. Furthermore, as shown in the example of the X^{2-} decay in a single QD in Section 4.1.3, the singlet and triplet spin structure of the *final* state explains in the case of a X^{2-} the observation of two emission lines.

Similarly as for the single QD spectra discussed in Section 3.6, the spectra of excitonic states in QDMs can be described when taking into account the direct

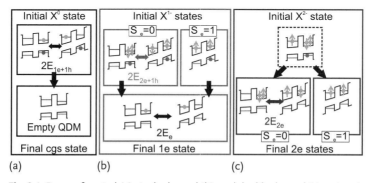

Fig. 8.4 Decay of neutral (a) singly charged (b), and doubly charged (c) exciton in a QDM.

and exchange Coulomb interactions, as well as quantum couplings for both the initial and final states of the radiative decay. In this section we present a detailed discussion of singly (trions) and doubly negatively charged excitons in a QDM. In a PL experiment the decays from the initial excitonic states X^0, X^{1-} and X^{2-}, were observed. The corresponding final states were the crystal ground state, the single-electron ($1e$) state, and the two-electron ($2e$) state, respectively. These transitions are summarized and shown schematically in Figure 8.4. In contrast to a single QD, the experimental data suggested here that electrons can tunnel between the two dots, whenever this is energetically favorable.

Here we mainly focus on the simplest case of the negative trion in a QDM and demonstrate the basic principle of charge and spin dependent couplings in both the initial and final state of the optical transition. For the negative trion, both Coulomb and Pauli blockade effects occur between the two electrons, in addition to quantum mechanical tunneling between the dots. Obviously, for the final $1e$ state, tunnel coupling also needs to be taken into account, while there is no Coulomb contribution present [293, 298].

In a first step, we calculate the absolute energies of the initial X^{1-} and the final $1e$ state of the optical transition. For these calculations we use a Hund–Mulliken approach, as outlined in Section 8.2, and a one-band effective mass Hamiltonian with harmonic confinement potentials for both electrons and holes [140]. From these calculations we are able to determine the spin and carrier number dependent energy states including their dependence on the electric field. The results are presented in Figure 8.5 with the spatial distribution of electrons and holes for the

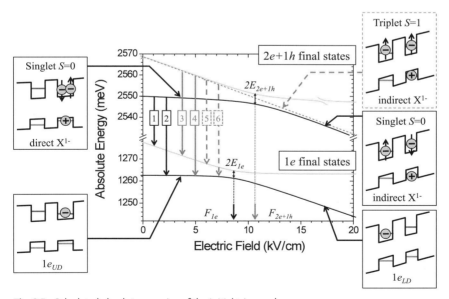

Fig. 8.5 Calculated absolute energies of the initial trion and final $1e$ state of the optical transition.

lowest lying trion and $1e$ levels. For both the initial trion and final electron state clear anticrossings, which are due to the tunnel coupling, are resolved. The resonance for the trion is shifted to higher electric field compared to the $1e$ due to the net attractive Coulomb interaction with the additional e–h-pair present in the QDM. Thus, to establish resonance this relative shift has to be overcome, giving rise to a detuning between the resonance fields F_{1e} and F_{2e+1h}.

Crucially, for X^{1-} the two-electron spin state has to be taken into account. These spins can form either an $S = 0$ singlet or $S = 1$ triplet wavefunction. For the triplet trion state no direct configuration is energetically accessible since the Pauli principle requires that for a symmetric spin wavefunction the two electrons must occupy different orbital states, when localized in the same dot. Therefore, for $S = 1$ only an indirect configuration exists, which shifts rapidly with F.

The absolute energies of all aforementioned states are plotted as a function of the externally applied electric field in Figure 8.5. For the $1e$ final state, a clear anticrossing between the weakly shifting electron level in the upper dot (e_{UD}) and the rapidly shifting state in the lower dot (e_{LD}) is expected. The minimum energy splitting $2E_{1e}$ observed at $F \sim 8.5\,$kV/cm reflects the tunnel coupling energy of a single electron. The singlet X^{1-} levels (solid lines in Figure 8.5) also show a resonance between weakly shifting direct and rapidly shifting indirect trion states, for which both electrons are either localized in the upper dot, or one electron is localized in each dot, respectively. As described before, this resonance is shifted in F relative to the $1e$ anticrossing. Moreover, a larger splitting $2E_{2e+1h} > 2E_{1e}$ is expected than for the single electron since two electrons mediate the coupling. In addition to the singlet trion levels, the indirect triplet state is plotted as a dashed line in Figure 8.5, which shifts rapidly with F and moves just as a straight line through the anticrossing singlet levels.

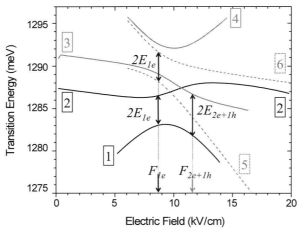

Fig. 8.6 Energies of the initial trion and final $1e$ state of the optical transition marked in Figure 8.5.

Obviously, the individual Coulomb interaction terms determine a distinct resonance field for a given number of carriers in the QDM. These characteristic offsets in F provide the potential to establish quantum mechanical coupling specifically for the initial or final state of certain optical transitions.

The energy of the optical transition observed in experiment is calculated as the energy difference of the initial trion and the final $1e$ state. The six possible transitions between the three trion and two electron states are marked and labeled in Figure 8.5 and their transition energies are plotted in Figure 8.6. From this characteristic X-pattern the initial and final state coupling energies can be determined. The transitions labeled as 1 and 2 (black solid lines) occur from a common initial singlet trion level into the delocalized $1e$ states and, therefore, measure $2E_{1e}$ at F_{1e}. In contrast, transitions 1 (black) and 3 (gray) have a common final state, and their minimum splitting $2E_{2e+1h}$ occurs at the singlet trion resonance at F_{2e+1h}. For the X^{1-} triplet (dashed lines), only one initial state exists, therefore, only the $1e$ resonance is mapped out, giving rise to a regular anticrossing pattern of transitions 5 and 6 at F_{1e}.

8.3.2
Electric Field Induced Coupling of Charged Trions in a QD Molecule

We have shown in Section 8.3.1 that the interplay of spin and charge dependent quantum coupling in the initial and final state of the optical transitions gives rise to a characteristic X-pattern of spectral lines. Figure 8.7 shows a grayscale plot of spectra recorded in an emission experiment from a single QDM as a function of F. In addition to the X^0 resonance at $F \sim 15.8$ kV/cm the anticipated X-pattern is clearly resolved. In particular, transition 2, which corresponds to the evolution of

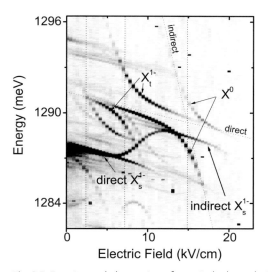

Fig. 8.7 Experimental observation of negatively charged trions in a tunable QDM.

a direct, singlet X^{1-} into an indirect configuration, is clearly resolved. In addition, the weaker features of transitions 1 and 3 are found. Moreover, the anticrossing of lines 5 and 6 originating from the decay of the indirect triplet trion state can be identified.

The extracted peak positions of the observed X^{1-} and X^{2-} emissions are summarized and plotted as symbols in Figure 8.8a and b, respectively. When comparing the experimental data for the trion emission to our calculations (shown as lines in Figure 8.8), we find excellent agreement between the calculated and experimentally observed transition energies. This strongly supports our approach using molecular theory for the states containing one and two delocalized electrons in the QDM, which are summarized in Figure 8.4. We further analyzed the splittings $2 E_m$ for the observed resonances at fields F_m in initial and final states by using Eq. (8.12). The results of these fits are shown in Figure 8.8. We find that the coupling energies of X^0 and $1e$ of 3.2 and 3.4 meV are very similar since in both cases one electron mediates the coupling. This finding confirms that tunnel coupling is the dominant mechanism in our system.

The shift of the resonance fields of the neutral exciton and the singlet trion, $F_{1e+1h} = 15.8\,\text{kV/cm}$ and $F_{2e+1h} = 10.6\,\text{kV/cm}$, of $\Delta F_{+1e} = -5.2\,\text{kV/cm}$ is determined by the change of the electrostatic energy upon adding one electron to the QDM. This shift can be analyzed further within a capacitive model, which is typically applied for electrostatic QDs [20, 86]. In this model the change of electrostatic

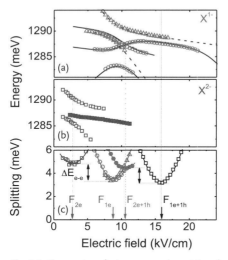

Fig. 8.8 Comparison between experiment (symbols) and theory (lines). (a) singlet (open circles) and triplet (open triangles) trions, (b) X^{2-} singlet (open squares) and triplet (full squares). (c) Measured splitting with fits of Eq. (8.12).

energy upon adding an extra carrier is given by

$$\Delta E_C = d\Delta F = e^2/C_M \left(\frac{C_1 C_2}{C_M^2} - 1 \right)^{-1}, \tag{8.16}$$

with d being the interdot separation, $C_{1,2}$ the single QD capacitance, and C_M the interdot mutual capacitance [86]. Using the measured value of $\Delta E_C = \Delta F_{+1e} \cdot d = 5.2\,\text{meV}$ and $C_1 = C_2 = 10\,\text{aF}$ we obtain a value of $C_M = 3\,\text{aF}$. The ratio of $C_M/C_{1,2} \sim 0.3$ confirms the relatively strong electrostatic interaction between the two dots.

Moreover, a similar shift of the critical field F_{crit} as observed between X^0 and X^{1-} is expected between the $1e$ and $2e$ electron state resonances. These two states are the final states of the trion and doubly charged X^{2-} transitions, respectively, as shown in Figure 8.4c. Analogous to the trion initial states, for the $2e$ final states the spin configuration has to be considered. As shown for X^{1-}, the singlet $2e$ state should show an anticrossing between a direct and an indirect configuration, in contrast to the triplet, which should not hybridize due to the large splitting to the p shell when compared to the $e - e$ Coulomb energy. The electron triplet should therefore remain indirect and cross through the singlet lines. From our electrostatic considerations presented for the X^0 and singlet X^{1-} resonance fields, we expect the $2e$ to be shifted by $\sim F_{+1e}$ with respect to the $1e$ resonance. In the spectra in Figure 8.7 a pair of anticrossing emission lines with a third line crossing through is observed at $F_{2e} \sim 2.8\,\text{kV/cm}$, very close to the expected value of $3.2\,\text{kV/cm}$. In the detailed analysis in Figure 8.8b and c of the doubly charged X^{2-} exciton, we find the same coupling energy $2E_{2e} = 4.4\,\text{meV}$ as for the trion singlet resonance, to which two electrons also contribute. This and the observation with a crossing triplet line provide further evidence for our assignment. We observe that the increased coupling energy for the resonances involving two instead of one electron is analogous to shifts in the spectrum of the neutral and ionized hydrogen molecule. This shift is given by the different Coulomb interactions when the two electrons are on the same dot or distributed among two dots, involving different direct and exchange Coulomb interaction terms. We can measure this splitting to be 1.25 meV in the splitting between the singlets X^{1-} and X^0 at a high electric field. Remarkably, this splitting agrees well with the observed change in coupling energy, underlining the analogy between an artificial QDM and a real hydrogen molecule. We want to underline here that this effect cannot be observed directly in a real hydrogen molecule since the antibonding orbitals are energetically above the vacuum level. Another remarkable effect that is not observable in an "atomic" molecule is the formation of an antibonding ground state. Recently, this counterintuitive phenomenon was observed in specially designed "artificial" molecules by Doty *et al.* [299]. Just taking these two examples, artificial molecules formed by two QDs provide a fascinating playground to test molecular physics. Moreover, new and surprising effects come into focus since tunable electronic and structural properties offer a unique toolkit for tailored artificial molecules.

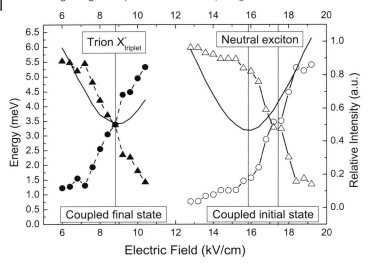

Fig. 8.9 Relative intensities (symbols) and splittings (lines) for trion-triplet and neutral exciton resonance demonstrating final and initial state couplings, respectively.

We have shown in Section 8.1 that for the neutral exciton anticrossing at $F = F_{1e+1h}$, the two emission lines exhibit the minimum splitting. However, a higher intensity is observed for the lower-energy branch, as can be seen in Figure 8.7. This finding is consistent with relaxation from the upper (antibonding) to the lower (bonding) exciton state within the radiative lifetime of the two states. Such relaxation can only occur if coupling is present in the optically excited, initial state of the transition. In contrast, we have shown that for the decay of the triplet trion state no coupling takes place in the initial state but only between the final $1e$ states, as highlighted in Figure 8.4. Therefore, we plot the energy splittings and relative intensities of the X^0 and triplet X^{1-} (transitions 5 and 6) anticrossings as a function of the electric field in Figure 8.9. In contrast to the neutral exciton, for the trion resonance the relative intensities are equal at the corresponding critical field F_{1e}. This finding clearly excludes relaxation and provides evidence that this transition occurs from a single initial state into a coupled final state, consistent with our previously developed picture. We stress that relaxation can lead to the suppression of transition 4 and the different intensities of the other singlet transitions (1–3). In addition, for the $2e$ anticrossing, no relaxation is observed as can be seen in Figure 8.7.

8.4
Future Directions

In contrast to single QDs, quantum dot molecules provide an inherently scalable architecture for the implementation of quantum logic using spin or charge excitations. In particular, $2e$ states are of interest for such purposes and schemes for optical spin gating have attracted much interest [300, 301]. Recently, the spin fine structure of singly and doubly charged QDMs as well as the biexciton states have been analyzed in great detail [295, 296, 302, 303]. Understanding the peculiarities of these states is essential for the realization of any related scheme for quantum information processing. At the time when this book was written, the latest developments in this field included conditional absorption experiments in coupled QDs [304] and the demonstration of optical spin initialization and nondestructive measurement in a QDM [242] using methods discussed in Chapter 7. Further directions for future research also include direct electron spin resonance experiments [261] for which a pronounced change of the g-factor at resonance in a QDM [294, 299, 305] may be exploited. Moreover, QDMs embedded in microcavities offer a large potential for more sophisticated schemes where spins interact via the optical field of a microcavity [181]. Since QDs can be precisely and controllably coupled in these systems [306], these techniques and methods might become crucial in the future. The first promising steps towards coupled and controllable cavity emitter coupling have already been taken [179, 307].

We hope that we have been able to give our readers a book at hand that not only provides an in-depth overview of the underlying experimental and theoretical concepts, but also shows the vast potential and depth of this field, which is at the forefront of contemporary solid state and nanoscale research.

Appendix A
Valence Band States for Spherical Confinement

We apply the usual basis of the band-edge Bloch states u_{jz} with angular momentum $j = 3/2$ in the following. In contrast to the conduction band states, the classification of the valence band states according to the total angular momentum $\mathbf{F}$ and the parity operator $\mathbf{P}$ provides coupling of l and $l + 2$ radial states (i.e., $s - d$ and $p - f$ coupling) in the envelope radial wave functions. We represent the valence band states as $|n L_F; F_z\rangle$. Applying the usual relations of Clebsch–Gordan coefficients, we obtain the $F = 3/2$ and $F = 1/2$ states with even parity as

$$\langle \mathbf{r}|1S_{3/2}; +3/2\rangle = \left(R_0 Y_{00} + \sqrt{\frac{1}{5}} R_2 Y_{20} \right) u_{+3/2}$$
$$- \sqrt{\frac{2}{5}} R_2 Y_{21} u_{+1/2} + \sqrt{\frac{2}{5}} R_2 Y_{22} u_{-1/2} , \tag{A1}$$

$$\langle \mathbf{r}|1S_{3/2}; +1/2\rangle = \sqrt{\frac{2}{5}} R_2 Y_{2,-1} u_{+3/2}$$
$$+ \left(R_0 Y_{00} - \sqrt{\frac{1}{5}} R_2 Y_{20} \right) u_{+1/2}$$
$$+ \sqrt{\frac{2}{5}} R_2 Y_{22} u_{-3/2} , \tag{A2}$$

$$\langle \mathbf{r}|1S_{3/2}; -1/2\rangle = \sqrt{\frac{2}{5}} R_2 Y_{2,-2} u_{+3/2}$$
$$+ \left(R_0 Y_{00} - \sqrt{\frac{1}{5}} R_2 Y_{20} \right) u_{-1/2}$$
$$+ \sqrt{\frac{2}{5}} R_2 Y_{21} u_{-3/2} , \tag{A3}$$

$$\langle \mathbf{r}|1S_{3/2}; -3/2\rangle = \sqrt{\frac{2}{5}} R_2 Y_{2,-2} u_{+1/2} - \sqrt{\frac{2}{5}} R_2 Y_{2,-1} u_{-1/2}$$
$$+ \left(R_0 Y_{00} + \sqrt{\frac{1}{5}} R_2 Y_{20} \right) u_{-3/2} , \tag{A4}$$

Spins in Optically Active Quantum Dots. Concepts and Methods.
Oliver Gywat, Hubert J. Krenner, and Jesse Berezovsky
Copyright © 2010 WILEY-VCH Verlag GmbH & Co. KGaA, Weinheim
ISBN: 978-3-527-40806-1

$$\langle \mathbf{r}|1S_{1/2}; +1/2\rangle = R_2 \left(-\sqrt{\frac{1}{10}}\, Y_{2,-1}\, u_{+3/2} + \sqrt{\frac{1}{5}}\, Y_{20}\, u_{+1/2} \right)$$

$$+ R_2 \left(-\sqrt{\frac{3}{10}}\, Y_{21}\, u_{-1/2} + \sqrt{\frac{2}{5}}\, Y_{22}\, u_{-3/2} \right), \tag{A5}$$

$$\langle \mathbf{r}|1S_{1/2}; -1/2\rangle = R_2 \left(-\sqrt{\frac{2}{5}}\, Y_{2,-2}\, u_{+3/2} + \sqrt{\frac{3}{10}}\, Y_{2,-1}\, u_{+1/2} \right)$$

$$+ R_2 \left(-\sqrt{\frac{1}{5}}\, Y_{20}\, u_{-1/2} + \sqrt{\frac{1}{10}}\, Y_{21}\, u_{-3/2} \right). \tag{A6}$$

Similarly, the $F = 3/2$ and $F = 1/2$ states with odd parity are obtained as

$$\langle \mathbf{r}|1P_{3/2}; +3/2\rangle = -\left(\sqrt{\frac{3}{5}}\, R_1 Y_{10} + \sqrt{\frac{1}{35}}\, R_3 Y_{30} \right) u_{+3/2}$$

$$+ \left(\sqrt{\frac{2}{5}}\, R_1 Y_{11} + 2\sqrt{\frac{1}{35}}\, R_3 Y_{31} \right) u_{+1/2}$$

$$- \sqrt{\frac{2}{7}}\, R_3 Y_{32}\, u_{-1/2} + 2\sqrt{\frac{1}{7}}\, R_3 Y_{33}\, u_{-3/2}, \tag{A7}$$

$$\langle \mathbf{r}|1P_{3/2}; +1/2\rangle = -\left(\sqrt{\frac{2}{5}}\, R_1 Y_{1,-1} + 2\sqrt{\frac{1}{35}}\, R_3 Y_{3,-1} \right) u_{+3/2}$$

$$+ \left(-\sqrt{\frac{1}{15}}\, R_1 Y_{10} + 3\sqrt{\frac{1}{35}}\, R_3 Y_{30} \right) u_{+1/2}$$

$$+ \left(2\sqrt{\frac{2}{15}}\, R_1 Y_{11} - 2\sqrt{\frac{3}{35}}\, R_3 Y_{31} \right) u_{-1/2}$$

$$+ \sqrt{\frac{2}{7}}\, R_3 Y_{32}\, u_{-3/2}, \tag{A8}$$

$$\langle \mathbf{r}|1P_{3/2}; -1/2\rangle = -\sqrt{\frac{2}{7}}\, R_3 Y_{3,-2}\, u_{+3/2}$$

$$+ \left(-2\sqrt{\frac{2}{15}}\, R_1 Y_{1,-1} + 2\sqrt{\frac{3}{35}}\, R_3 Y_{3,-1} \right) u_{+1/2}$$

$$+ \left(\sqrt{\frac{1}{15}}\, R_1 Y_{10} - 3\sqrt{\frac{1}{35}}\, R_3 Y_{30} \right) u_{-1/2}$$

$$+ \left(\sqrt{\frac{2}{5}}\, R_1 Y_{11} + 2\sqrt{\frac{1}{35}}\, R_3 Y_{31} \right) u_{-3/2}, \tag{A9}$$

$$\langle \mathbf{r}|1P_{3/2};-3/2\rangle = -2\sqrt{\frac{1}{7}}R_3\,Y_{3,-3}\,u_{+3/2} + \sqrt{\frac{2}{7}}R_3\,Y_{3,-2}\,u_{+1/2}$$

$$-\left(\sqrt{\frac{2}{5}}R_1\,Y_{1,-1} + 2\sqrt{\frac{1}{35}}R_3\,Y_{3,-1}\right)u_{-1/2}$$

$$+\left(\sqrt{\frac{3}{5}}R_1\,Y_{10} + \sqrt{\frac{1}{35}}R_3\,Y_{30}\right)u_{-3/2}\,, \tag{A10}$$

$$\langle \mathbf{r}|1P_{1/2};+1/2\rangle = R_1\left(-\sqrt{\frac{1}{2}}Y_{1,-1}\,u_{+3/2} - \sqrt{\frac{1}{3}}Y_{10}\,u_{+1/2}\right)$$

$$+ R_1\sqrt{\frac{1}{6}}Y_{11}\,u_{-1/2}\,, \tag{A11}$$

$$\langle \mathbf{r}|1P_{1/2};-1/2\rangle = R_1\left(\sqrt{\frac{1}{6}}Y_{1,-1}\,u_{+1/2} - \sqrt{\frac{1}{3}}Y_{10}\,u_{-1/2}\right)$$

$$+ R_1\sqrt{\frac{1}{2}}Y_{11}\,u_{-3/2}\,. \tag{A12}$$

Appendix B
List of Constants

Table B.1 Values of important constants recommended by the Committee on Data for Science and Technology (CODATA) [308]. See also http://physics.nist.gov/cuu/Constants/index.html.

Quantity	Symbol	Value	Unit
Speed of light	c	299 792 458	m/s
Magnetic constant	μ_0	$4\pi \cdot 10^7 = 12.566\,370\,614\ldots \cdot 10^{-7}$	N/A^2
Electric constant	$\epsilon_0 = 1/\mu_0 c^2$	$8.854\,187\,817 \cdot 10^{-12}$	F/m
Impedance of the vacuum	$Z_0 = \sqrt{\mu_0/\epsilon_0}$	$376.730\,313\,461\ldots$	Ω
Elementary charge	e	$1.602\,176\,487(40) \cdot 10^{-19}$	A s
Free electron mass	m_e	$9.109\,382\,15(45) \cdot 10^{-31}$	kg
	$m_e c^2$	$510\,998.910(13)$	eV
Planck's constant	h	$6.626\,068\,96(33) \cdot 10^{-34}$	J s
		$4.135\,667\,33(10) \cdot 10^{-15}$	eV s
Reduced Planck's constant	$\hbar = h/2\pi$	$1.054\,571\,628(53) \cdot 10^{-34}$	J s
		$6.582\,118\,99(16) \cdot 10^{-16}$	eV s
Magnetic flux quantum	$\Phi_0 = h/2e$	$2.067\,833\,667(52) \cdot 10^{-15}$	Wb
Conductance quantum	$G_0 = 2e^2/h$	$7.748\,091\,7004(53) \cdot 10^{-5}$	S
von Klitzing constant	$R_K = h/e^2$	$25\,812.807\,557(18)$	Ω
Fine structure constant	$\alpha = e^2/4\pi\epsilon_0\hbar c$	$7.297\,352\,5376(50) \cdot 10^{-3}$	
Inverse fine structure constant	α^{-1}	$1/137.035\,999\,711(96)$	
Bohr radius	$a_B = 4\pi\epsilon_0\hbar^2/m_e e^2$	$0.529\,177\,208\,59(36) \cdot 10^{-10}$	m
Rydberg energy	$R_\infty = \alpha^2 m_e c/2h$	$10\,973\,731.568\,527(73)$	1/m
	$R_\infty hc$	$13.605\,691\,93(34)$	eV
Bohr's magneton	$\mu_B = e\hbar/2m_e$	$927.400\,915(23) \cdot 10^{-26}$	J/T
		$5.788\,381\,7555(79) \cdot 10^{-5}$	eV/T
Free electron g-factor	g	$-2.002\,319\,304\,3622(15)$	
Free electron magnetic moment	μ_e	$-928.476\,377(23) \cdot 10^{-26}$	J/T
	μ_e/μ_B	$-1.001\,159\,652\,181\,11(74)$	
Boltzmann constant	k_B	$1.380\,6504(24) \cdot 10^{-23}$	J/K
		$8.617\,343(15) \cdot 10^{-5}$	eV/K
electron volt e/C	eV	$1.602\,176\,487(40) \cdot 10^{-19}$	J

Spins in Optically Active Quantum Dots. Concepts and Methods.
Oliver Gywat, Hubert J. Krenner, and Jesse Berezovsky
Copyright © 2010 WILEY-VCH Verlag GmbH & Co. KGaA, Weinheim
ISBN: 978-3-527-40806-1

Appendix C
Material Parameters

Table C.1 Parameters of III–V and II–VI semiconductors [50, 309–314].

	GaAs	AlAs	InAs	ZnSe	CdSe	CdS
Lattice constant						
Zincblende a (nm) – 300 K	0.5653	0.5660	0.6058	0.5668	–	0.5818
Wurtzite a axis (nm)					0.423	0.414
Wurtzite c axis					0.7011	0.671
Bandgap – 0 K (eV)	1.52	2.23	0.42	2.820	1.841	2.585
300 K (eV)	1.42	2.15	0.35	2.713	1.751	2.485
E_{gap}^{Γ}		3.02				
Band minimum	Γ	X	Γ	Γ	Γ	Γ
Effective mass (m_0)						
m_e^*	0.067	0.15 (Γ)	0.026	0.15	0.11	0.21
$m_{(e,l)}^*$	1.3 (L)	1.1				
$m_{(e,t)}^*$	0.23 (L)	0.19				
m_{hh}^*	0.5	0.5	0.41	1.4	> 1 ∥ c axis	0.64–0.685
m_{lh}^*	0.082	0.15	0.026	0.15	0.45 ⊥ c axis	
Spin-orbit split off energy						
Δ_{so} (eV)	0.34	0.28	0.38	0.40	0.42	0.062
Electron g-factor	−0.44	1.85 (L)	−14.8	1.15	0.68	
Dielectric constant ϵ_r	13.2	10.1	15.1	9.1	10.16 (∥)	9.38
					9.29 (⊥)	
Kane energy E_P (eV)	25.7	21.1	22.2		17.5	

Spins in Optically Active Quantum Dots. Concepts and Methods.
Oliver Gywat, Hubert J. Krenner, and Jesse Berezovsky
Copyright © 2010 WILEY-VCH Verlag GmbH & Co. KGaA, Weinheim
ISBN: 978-3-527-40806-1

References

1 Sakurai, J. J. (1995) *Modern Quantum Mechanics, Revised Edition*. Addison-Wesley Publishing Company, Reading, Massachusetts.

2 Hanson, R., Dobrovitski, V. V., Feiguin, A. E., Gywat, O., and Awschalom, D. D. (2008) Coherent dynamics of a single spin interacting with an adjustable spin bath. *Science*, **320**, 352–355.

3 Ashoori, R. C. (1996) Electrons in artifical atoms. *Nature (London)*, **379**, 413.

4 Leo, W. (1926) Über ausgewählte Gebiete des Heliumspektrums. *Annalen der Physik*, **386**, 757–799.

5 Preskill, J. Course information for physics 219/computer science 219 quantum computation. http://www.theory.caltech.edu/preskill/ph229/.

6 Shor, P. W. (1997) Polynomial-time algorithms for prime factorization and discrete logarithms on a quantum computer. *SIAM Journal on Scientific and Statistical Computing*, **26**, 1484.

7 Grover, L. K. (1997) Quantum mechanics helps in searching for a needle in a haystack. *Physical Review Letters*, **79**, 325–328.

8 DiVincenzo, D. P. (2000) The physical implementation of quantum computation. *Fortschritte der Physik*, **48**, 771–784.

9 Barenco, A., Bennett, C. H., Cleve, R., DiVincenzo, D. P., Margolus, N., Shor, P., Sleator, T., Smolin, J. A., and Weinfurter, H. (1995) Elementary gates for quantum computation. *Physical Review A*, **52**, 3457–3467.

10 DiVincenzo, D. P. (1995) Two-bit gates are universal for quantum computation. *Physical Review A*, **51**, 1015–1022.

11 Cerletti, V., Coish, W. A., Gywat, O., and Loss, D. (2005) Recipes for spin-based quantum computing. *Nanotechnology*, **16**, R27–R49.

12 Nielsen, M. A. and Chuang, I. L. (2000) *Quantum Computation and and Quantum Information*. Cambridge University Press, Cambridge.

13 Waks, E., Inoue, K., Santori, C., Fattal, D., Vuckovic, J., Solomon, G. S., and Yamamoto, Y. (2002) Secure communication: quantum cryptography with a photon turnstile. *Nature (London)*, **420**, 762.

14 Fattal, D., Diamanti, E., Inoue, K., and Yamamoto, Y. (2004) Quantum teleportation with a quantum dot single photon source. *Physical Review Letters*, **92**, 037904.

15 Loss, D. and DiVincenzo, D. P. (1998) Quantum computation with quantum dots. *Physical Review A*, **57**, 120–126.

16 Openov, L. A. (1999) Resonant electron transfer between quantum dots. *Physical Review B*, **60**, 8798–8803.

17 Biolatti, E., Iotti, R. C., Zanardi, P., and Rossi, F. (2000) Quantum information processing with semiconductor macroatoms. *Physical Review Letters*, **85**, 5647–5650.

18 Troiani, F., Molinari, E., and Hohenester, U. (2003) High-finesse optical quantum gates for electron spins in artificial molecules. *Physical Review Letters*, **90**, 206802.

19 Lovett, B. W., Reina, J. H., Nazir, A., and Briggs, G. A. D. (2003) Optical schemes for quantum computation in quantum dot molecules. *Physical Review B*, **68**, 205319.

Spins in Optically Active Quantum Dots. Concepts and Methods.
Oliver Gywat, Hubert J. Krenner, and Jesse Berezovsky
Copyright © 2010 WILEY-VCH Verlag GmbH & Co. KGaA, Weinheim
ISBN: 978-3-527-40806-1

20 Hanson, R., Kouwenhoven, L. P., Petta, J. R., Tarucha, S., and Vandersypen, L. M. K. (2007) Spins in few-electron quantum dots. *Reviews of Modern Physics*, **79**, 1217.

21 Brunner, K., Bockelmann, U., Abstreiter, G., Walther, M., Böhm, G., Tränkle, G., and Weimann, G. (1992) Photoluminescence from a single GaAs/AlGaAs quantum dot. *Physical Review Letters*, **69**, 3216–3219.

22 Brunner, K., Abstreiter, G., Böhm, G., Tränkle, G., and Weimann, G. (1994) Sharp-line photoluminescence and two-photon absorption of zero-dimensional biexcitons in a GaAs/AlGaAs structure. *Physical Review Letters*, **73**, 1138–1141.

23 Zrenner, A., Butov, L. V., Hagn, M., Abstreiter, G., Böhm, G., and Weimann, G. (1994) Quantum dots formed by interface fluctuations in AlAs/GaAs coupled quantum well structures. *Physical Review Letters*, **72**, 3382–3385.

24 Gammon, D., Snow, E. S., Shanabrook, B. V., Katzer, D. S., and Park, D. (1996) Fine structure splitting in the optical spectra of single GaAs quantum dots. *Physical Review Letters*, **76**, 3005–3008.

25 Guest, J. R., Stievater, T. H., Chen, G., Tabak, E. A., Orr, B. G., Steel, D. G., Gammon, D., and Katzer, D. S. (2001) Near-field coherent spectroscopy and microscopy of a quantum dot system. *Science*, **293**, 2224–2227.

26 Bonadeo, N. H., Erland, J., Gammon, D., Park, D., Katzer, D. S., and Steel, D. G. (1998) Coherent optical control of the quantum state of a single quantum dot. *Science*, **282**, 1473–1476.

27 Stievater, T. H., Li, X., Steel, D. G., Gammon, D., Katzer, D. S., Park, D., Piermarocchi, C., and Sham, L. J. (2001) Rabi oscillations of excitons in single quantum dots. *Physical Review Letters*, **87**, 133603.

28 Chen, G., Stievater, T. H., Batteh, E. T., Li, X., Steel, D. G., Gammon, D., Katzer, D. S., Park, D., and Sham, L. J. (2002) Biexciton quantum coherence in a single quantum dot. *Physical Review Letters*, **88**, 117901.

29 Li, X., Wu, Y., Steel, D., Gammon, D., Stievater, T. H., Katzer, D. S., Park, D., Piermarocchi, C., and Sham, L. J. (2003) An all-optical quantum gate in a semiconductor quantum dot. *Science*, **301**, 809–811.

30 Stranski, I. N. and Krastanow, L. (1937) Zur Theorie der orientierten Ausscheidung von Ionenkristallen aufeinander. *Monatshefte für Chemie/Chemical Monthly*, **71**, 351–364.

31 Goldstein, L., Glas, F., Marzin, J. Y., Charasse, M. N., and Roux, G. L. (1985) Growth by molecular beam epitaxy and characterization of InAs/GaAs strained-layer superlattices. *Applied Physics Letters*, **47**, 1099–1101.

32 Eaglesham, D. J. and Cerullo, M. (1990) Dislocation-free Stranski-Krastanow growth of Ge on Si(100). *Physical Review Letters*, **64**, 1943–1946.

33 Leonard, D., Krishnamurthy, M., Reaves, C. M., Denbaars, S. P., and Petroff, P. M. (1993) Direct formation of quantum-sized dots from uniform coherent islands of InGaAs on GaAs surfaces. *Applied Physics Letters*, **63**, 3203–3205.

34 Leonard, D., Pond, K., and Petroff, P. M. (1994) Critical layer thickness for self-assembled InAs islands on GaAs. *Physical Review B*, **50**, 11687–11692.

35 Marzin, J.-Y., Gerard, J.-M., Izrael, A., Barrier, D., and Bastard, G. (1994) Photoluminescence of single InAs quantum dots obtained by self-organized growth on GaAs. *Physical Review Letters*, **73**, 716–719.

36 Grundmann, M. *et al.* (1995) Ultranarrow luminescence lines from single quantum dots. *Physical Review Letters*, **74**, 4043–4046.

37 Dekel, E., Gershoni, D., Ehrenfreund, E., Spektor, D., Garcia, J. M., and Petroff, P. M. (1998) Multiexciton spectroscopy of a single self-assembled quantum dot. *Physical Review Letters*, **80**, 4991–4994.

38 Landin, L., Miller, M. S., Pistol, M. E., Pryor, C. E., and Samuelson, L. (1998) Optical studies of individual InAs quantum dots in GaAs: Few particle effects. *Science*, **280**, 262–264.

39 Bayer, M., Stern, O., Hawrylak, P., Fafard, S., and Forchel, A. (2000) Hidden symmetries in the energy levels of excitonic 'artificial atoms'. *Nature (London)*, **405**, 923–926.

40 Warburton, R. J., Schäflein, C., Haft, D., Bickel, F., Lorke, A., Karrai, K., Garcia, J. M., Schoenfeld, W., and Petroff, P. M. (2000) Optical emission from a charge-tunable quantum ring. *Nature (London)*, **405**, 926–929.

41 Kamada, H., Gotoh, H., Temmyo, J., Takagahara, T., and Ando, H. (2001) Exciton Rabi oscillation in a single quantum dot. *Physical Review Letters*, **87**, 246401.

42 Zrenner, A., Beham, E., Stufler, S., Findeis, F., Bichler, M., and Abstreiter, G. (2002) Coherent properties of a two-level system based on a quantum-dot photodiode. *Nature (London)*, **418**, 612–614.

43 Besombes, L., Baumberg, J. J., and Motohisa, J. (2003) Coherent spectroscopy of optically gated charged single InGaAs quantum dots. *Physical Review Letters*, **90**, 257402.

44 Xu, X., Sun, B., Berman, P. R., Steel, D. G., Bracker, A. S., Gammon, D., and Sham, L. J. (2007) Coherent optical spectroscopy of a strongly driven quantum dot. *Science*, **317**, 929.

45 Jundt, G., Robledo, L., Högele, A., Fält, S., and Imamoğlu, A. (2008) Observation of dressed excitonic states in a single quantum dot. *Physical Review Letters*, **100**, 177401.

46 Klitzing, K. v., Dorda, G., and Pepper, M. (1980) New method for high-accuracy determination of the fine-structure constant based on quantized hall resistance. *Physical Review Letters*, **45**, 494–497.

47 Tsui, D. C., Stormer, H. L., and Gossard, A. C. (1982) Two-dimensional magneto-transport in the extreme quantum limit. *Physical Review Letters*, **48**, 1559–1562.

48 Miller, D. A. B., Chemla, D. S., Damen, T. C., Gossard, A. C., Wiegmann, W., Wood, T. H., and Burrus, C. A. (1985) Electric field dependence of optical absorption near the band gap of quantum-well structures. *Physical Review B*, **32**, 1043–1060.

49 Esaki, L. (1992) Do-it-yourself quantum mechanics in low-dimensional structures. *Physica Scripta*, **T42**, 102–109.

50 Davies, J. H. (2006) *The Physics of Low-dimensional Semiconductors*. Cambridge University Press, Cambridge.

51 Faist, J., Capasso, F., Sivco, D. L., Sirtori, C., Hutchinson, A. L., and Cho, A. Y. (1994) Quantum cascade laser. *Science*, **264**, 553–556.

52 Capasso, F. (1987) Band-gap engineering – From physics and materials to new semiconductor devices. *Science*, **235**, 172–176.

53 Capasso, F., Faist, J., and Sirtori, C. (1996) Mesoscopic phenomena in semiconductor nanostructures by quantum design. *Journal of Mathematical Physics*, **37**, 4775–4792.

54 Herman, M. A., Richter, W., and Sitter, H. (2004) *Epitaxy – Physical principles and technical implementations*, vol. 62 of *Springer Series in Material Science*. Springer, Heidelberg.

55 Shchukin, V. A., Ledentsov, N. N., and Bimberg, D. (eds) (2004) *Epitaxy of Nanostructures*. Nanoscience and Technology, Springer, Heidelberg.

56 Yu, P. Y. and Cardona, M. (2005) *Fundamentals of Semiconductors: Physics and Material properties*. Springer, Berlin.

57 Frank, F. C. and van der Merwe, J. H. (1949) One-dimensional dislocations. i. static theory. *Proceedings of the Royal Society*, **A 198**, 205–216.

58 Volmer, M. and Weber, A. (1926) Nucleus formation in supersaturated systems. *Zeitschrift für physikalische Chemie*, **119**, 277–301.

59 Brunner, K., Abstreiter, G., Böhm, G., Tränkle, G., and Weimann, G. (1994) Sharp-line photoluminescence of excitons localized at GaAs/AlGaAs quantum well inhomogeneities. *Applied Physics Letters*, **64**, 3320–3322.

60 Berezovsky, J., Mikkelsen, M. H., Gywat, O., Stoltz, N. G., Coldren, L. A., and Awschalom, D. D. (2006) Nondestructive optical measurements of a single electron spin in a quantum dot. *Science*, **314**, 1916.

61 Liu, N., Tersoff, J., Baklenov, O., A. L. Holmes, J., and Shih, C. K. (2000) Nonuniform composition profile in $In_{0.5}Ga_{0.5}As$ alloy quantum dots. *Physical Review Letters*, **84**, 334–337.

62 Walther, T., Cullis, A. G., Norris, D. J., and Hopkinson, M. (2001) Nature of the Stranski–Krastanow transition during epitaxy of InGaAs on GaAs. *Physical Review Letters*, **86**, 2381–2384.

63 Xie, Q., Madhukar, A., Chen, P., and Kobayashi, N. P. (1995) Vertically self-organized InAs quantum box islands on GaAs(100). *Physical Review Letters*, **75**, 2542–2545.

64 Stangl, J., Holy, V., and Bauer, G. (2004) Structural properties of self-organized semiconductor nanostructures. *Reviews of Modern Physics*, **76**, 725.

65 Schmidt, O. G. (ed.) (2007) *Lateral Alignment of Epitaxial Quantum Dots.* Nanoscience and Technology, Springer, Heidelberg.

66 Bimberg, D. (ed.) (2008) *Semiconductor Nanostructures.* Nanoscience and Technology, Springer, Heidelberg.

67 Bruls, D. M., Koenraad, P. M., Salemink, H. W. M., Wolter, J. H., Hopkinson, M., and Skolnick, M. S. (2003) Stacked low-growth-rate InAs quantum dots studied at the atomic level by cross-sectional scanning tunneling microscopy. *Applied Physics Letters*, **82**, 3758–3760.

68 Wasilewski, Z. R., Fafard, S., and McCaffrey, J. P. (1999) Size and shape engineering of vertically stacked self-assembled quantum dots. *Journal of Crystal Growth*, **201/202**, 1131–1135.

69 Gerardot, B., Shtrichman, I., Hebert, D., and Petroff, P. M. (2003) Tuning of the electronic levels in vertically stacked InAs/GaAs quantum dots using crystal growth kinetics. *Journal of Crystal Growth*, **252**, 44–50.

70 Krenner, H. J., Stufler, S., Sabathil, M., Clark, E. C., Ester, P., Bichler, M., Abstreiter, G., Finley, J. J., and Zrenner, A. (2005) Recent advances in exciton-based quantum information processing in quantum dot nanostructures. *New Journal of Physics*, **7**, 184.

71 Bracker, A. S., Scheibner, M., Doty, M. F., Stinaff, E. A., Ponomarev, I. V., Kim, J. C., Whitman, L. J., Reinecke, T. L., and Gammon, D. (2006) Engineering electron and hole tunneling with asymmetric InAs quantum dot molecules. *Applied Physics Letters*, **89**, 233110.

72 Schedelbeck, G., Wegscheider, W., Bichler, M., and Abstreiter, G. (1997) Coupled quantum dots fabricated by cleaved edge overgrowth: From artificial atoms to molecules. *Science*, **278**, 1792–1795.

73 Solomon, G. S., Trezza, J. A., Marshall, A. F., and Harris, J. S., Jr. (1996) Vertically aligned and electronically coupled growth induced InAs islands in GaAs. *Physical Review Letters*, **76**, 952–955.

74 Bayer, M., Hawrylak, P., Hinzer, K., Fafard, S., Korkusinski, M., Wasilwski, Z. R., Stern, O., and Forchel, A. (2001) Coupling and entangling of quantum states in quantum dot molecules. *Science*, **291**, 451–453.

75 Ortner, G., Bayer, M., Larionov, A., Timofeev, V. B., Forchel, A., Lyanda-Geller, Y. B., Reinecke, T. L., Hawrylak, P., Fafard, S., and Wasilewski, Z. (2003) Fine structure of excitons in InAs/GaAs coupled quantum dots: A sensitive test of electronic coupling. *Physical Review Letters*, **90**, 086404.

76 Ortner, G., Bayer, M., Lyanda-Geller, Y., Reinecke, T. L., Kress, A., Reithmaier, J. P., and Forchel, A. (2005) Control of vertically coupled InGaAs/GaAs quantum dots with electric fields. *Physical Review Letters*, **94**, 157401.

77 Krenner, H. J., Sabathil, M., Clark, E. C., Kress, A., Schuh, D., Bichler, M., Abstreiter, G., and Finley, J. J. (2005) Direct observation of controlled coupling in an individual quantum dot molecule. *Physical Review Letters*, **94**, 057402.

78 Borri, P., Langbein, W., Woggon, U., Schwab, M., Bayer, M., Fafard, S., Wasilewski, Z., and Hawrylak, P. (2003) Exciton dephasing in quantum dot molecules. *Physical Review Letters*, **91**, 267401.

79 He, J., Krenner, H. J., Pryor, C., Zhang, J. P., Wu, Y., Allen, D. G., Morris, C. M., Sherwin, M. S., and Petroff, P. M. (2007) Growth, structural and optical properties of self-assembled (In,Ga)As Quantum Posts on GaAs. *Nano Letters*, **7**, 802–806.

80 Li, L. H., Ridha, P., Patriarche, G., Chauvin, N., and Fiore, A. (2008) Shape-engineered epitaxial InGaAs quantum rods for laser applications. *Applied Physics Letters*, **92**, 121102.

81 Wojs, A. and Hawrylak, P. (1995) Negatively charged magnetoexcitons in quan-

tum dots. *Physical Review B*, **51**, 10880–10885.

82 Hawrylak, P. (1999) Excitonic artificial atoms: Engineering optical properties of quantum dots. *Physical Review B*, **60**, 5597–5608.

83 Warburton, R. J., Miller, B. T., Durr, C. S., Bodefeld, C., Karrai, K., Kotthaus, J. P., Medeiros-Ribeiro, G., Petroff, P. M., and Huant, S. (1998) Coulomb interactions in small charge-tunable quantum dots: A simple model. *Physical Review B*, **58**, 16221–16231.

84 Shaji, N. *et al.* (2008) Spin blockade and lifetime-enhanced transport in a few-electron Si/SiGe double quantum dot. *Nature Physics*, **4**, 540–544.

85 Sailer, J. *et al.* (2009) A Schottky top-gated two-dimensional electron system in a nuclear spin free Si/SiGe heterostructure. *physica status solidi (RLL)*, **3**, 61–63.

86 van der Wiel, W. G., Franceschi, S. D., Elzerman, J. M., Fujisawa, T., Tarucha, S., and Kouwenhoven, L. P. (2003) Electron transport through double quantum dots. *Reviews of Modern Physics*, **75**, 1.

87 Trauzettel, B., Bulaev, D. V., Loss, D., and Burkard, G. (2007) Spin qubits in graphene quantum dots. *Nature Physics*, **3**, 192–196.

88 Stampfer, C., Schurtenberger, E., Molitor, F., Güttinger, J., Ihn, T., and Ensslin, K. (2008) Tunable graphene single electron transistor. *Nano Letters*, **8**, 2378–2383.

89 Neto, A. H. C., Guinea, F., Peres, N. M. R., Novoselov, K. S., and Geim, A. K. (2009) The electronic properties of graphene. *Reviews of Modern Physics*, **81**, 109.

90 Wegscheider, W., Schedelbeck, G., Abstreiter, G., Rother, M., and Bichler, M. (1997) Atomically precise GaAs/AlGaAs quantum dots fabricated by twofold cleaved edge overgrowth. *Physical Review Letters*, **79**, 1917–1920.

91 Wegscheider, W., Schedelbeck, G., Bichler, M., and Abstreiter, G. (1999) Atomically precise, coupled quantum dots fabricated by cleaved edge overgrowth. Kramer, B. (ed.), *Advances in Solid State Physics*, vol. 38, pp. 153–165, Springer, Berlin.

92 Hartmann, A., Ducommun, Y., Leifer, K., and Kapon, E. (1999) Structure and opti-

cal properties of semiconductor quantum nanostructures self-formed in inverted tetrahedral pyramids. *Journal of Physics Condensed Matter*, **11**, 5901–5915.

93 Zhu, Q., Karlsson, K. F., Byszewski, M., Rudra, A., Pelucchi, E., He, Z., and Kapon, E. (2009) Hybridization of electron and hole states in semiconductor quantum-dot molecules. *Small*, **5**, 329–335.

94 Hartmann, A., Ducommun, Y., Kapon, E., Hohenester, U., and Molinari, E. (2000) Few-particle effects in semiconductor quantum dots: Observation of multi-charged excitons. *Physical Review Letters*, **84**, 5648–5651.

95 Samuelson, L. *et al.* (2004) Semiconductor nanowires for 0d and 1d physics and applications. *Physica E*, **25**, 313–318, proceedings of the 13th International Winterschool on New Developments in Solid State Physics – Low-Dimensional Systems.

96 Thelander, C. *et al.* (2006) Nanowire-based one-dimensional electronics. *Materials Today*, **9**, 28–35.

97 Martensson, T., Svensson, C. P. T., Wacaser, B. A., Larsson, M. W., Seifert, W., Deppert, K., Gustafsson, A., Wallenberg, L. R., and Samuelson, L. (2004) Epitaxial III–V nanowires on silicon. *Nano Letters*, **4**, 1987–1990.

98 Lu, W. and Lieber, C. M. (2006) Semiconductor nanowires. *Journal of Physics D Applied Physics*, **39**, 387.

99 Panev, N., Persson, A. I., Sköld, N., and Samuelson, L. (2003) Sharp exciton emission from single InAs quantum dots in GaAs nanowires. *Applied Physics Letters*, **83**, 2238–2240.

100 Bleszynski-Jayich, A. C., Fröberg, L. E., Björk, M. T., Trodahl, H. J., Samuelson, L., and Westervelt, R. M. (2008) Imaging a one-electron InAs quantum dot in an InAs/InP nanowire. *Physical Review B*, **77**, 245327.

101 Mason, N., Biercuk, M. J., and Marcus, C. M. (2004) Local gate control of a carbon nanotube double quantum dot. *Science*, **303**, 655–658.

102 Jørgensen, H. I., Grove-Rasmussen, K., Wang, K.-Y., Blackburn, A. M., Flensberg, K., Lindelof, P. E., and Williams, D. A. (2008) Singlet–triplet physics and shell filling in carbon nanotube double quantum dots. *Nature Physics*, **4**, 536–539.

103 Pfund, A., Shorubalko, I., Ensslin, K., and Leturcq, R. (2007) Suppression of Spin Relaxation in an InAs Nanowire Double Quantum Dot. *Physical Review Letters*, **99**, 036801.

104 Brus, L. E. (1984) Electron–electron and electron–hole interactions in small semiconductor crystallites – the size dependence of the lowest excited electronic state. *Journal of Chemical Physics*, **80**, 4403–4409.

105 Murray, C. B., Norris, D. J., and Bawendi, M. (1993) Synthesis and characterization of nearly monodisperse CdE (E = S, Se, Te) semiconductor nanocrystallites. *Journal of the American Chemical Society*, **115**, 8706–8715.

106 Efros, Al. L. and Rosen, M. (2000) The electronic structure of semiconductor nanocrystals. *Annual Review of Materials Science*, **30**, 475–521.

107 Manna, L., Scher, E. C., and Alivisatos, A. P. (2000) Synthesis of soluble and processable rod-, arrow-, teardrop-, and tetrapod-shaped CdSe nanocrystals. *Journal of the American Chemical Society*, **122**, 12700–12706.

108 Hines, M. A. and Guyot-Sionnest, P. (1996) Synthesis and characterization of strongly luminescing ZnS-capped CdSe nanocrystals. *Journal of Physical Chemistry*, **100**, 468–471.

109 Dabbousi, B. O., Rodriguez-Viejo, J., Mikulec, F. V., Heine, J. R., Mattoussi, H., Ober, R., Jensen, K. F., and Bawendi, M. G. (1997) (CdSe)ZnS core-shell quantum dots: Synthesis and characterization of a size series of highly luminescent nanocrystallites. *Journal of Physical Chemistry B*, **101**, 9463–9475.

110 Mews, A., Eychmüller, A., Giersig, M., Schooss, D., and Weller, H. (1994) Preparation, characterization, and photophysics of the quantum-dot quantum-well system CdS/HgS/CdS. *Journal of Physical Chemistry*, **98**, 934–941.

111 Little, R. B., El-Sayed, M. A., Bryant, G. W., and Burke, S. (2001) Formation of quantum-dot quantum-well hetero-nanostructures with large lattice mismatch: ZnS/CdS/ZnS. *Journal of Chemical Physics*, **114**, 1813–1822.

112 Battaglia, D., Li, J. J., Wang, Y. J., and Peng, X. G. (2003) Colloidal two-dimensional systems: CdSe quantum shells and wells. *Angewandte Chemie – International Edition*, **42**, 5035–5039.

113 Efros, Al., Rosen, M., Kuno, M., Nirmal, M., Norris, D., and Bawendi, M. (1996) Band-edge exciton in quantum dots of semiconductors with a degenerate valence band: Dark and bright exciton states. *Physical Review B*, **54**, 4843–4856.

114 Norris, D. J., Sacra, A., Murray, C. B., and Bawendi, M. G. (1994) Measurement of the size-dependent hole spectrum in CdSe quantum dots. *Physical Review Letters*, **72**, 2612–2615.

115 Stern, N. P., Poggio, M., Bartl, M. H., Hu, E. L., Stucky, G. D., and Awschalom, D. D. (2005) Spin dynamics in electrochemically charged CdSe quantum dots. *Physical Review B*, **72**, 161303.

116 Schaller, R. and Klimov, V. (2004) High efficiency carrier multiplication in PbSe nanocrystals: Implications for solar energy conversion. *Physical Review Letters*, **92**, 186601.

117 Ashcroft, N. W. and Mermin, N. D. (1976) *Solid State Physics*. Saunders College Publishing, Philadelphia.

118 Madelung, O. (1996) *Introduction to Solid State Theory*, vol. 2 of *Springer Series in Solid State Sciences*. Springer, Berlin.

119 Kittel, C. (1988) *Einführung in die Festkörperphysik*. R. Oldenbourg Verlag, München.

120 Efros, Al. L. (1992) Luminescence polarization of CdSe microcrystals. *Physical Review B*, **46**, 7448–7458.

121 D'yakonov, M. I. and Perel', V. I. (1984) *Optical Orientation*, vol. 8 of *Modern Problems in Condensed Matter Sciences*. North Holland, Amsterdam.

122 Rashba, E. I. (1960) *Soviet Physics – Solid State (english translation)*, **2**, 1109.

123 Dresselhaus, G. (1955) Spin-orbit coupling effects in zinc blende structures. *Physical Review*, **100**, 580–586.

124 Ivchenko, E. I. (1995) *Superlattices and other heterostructures*. Springer, Berlin.

125 Landolt-Börnstein (1999) *Numerical Data and Functional Relationships in Science and*

Technology, vol. 41, Subvolume B of *New Series*. Springer, Heidelberg.

126 Zwerdling, S., Lax, B., Roth, L. M., and Button, K. J. (1959) Exciton and Magneto-Absorption of the Direct and Indirect Transitions in Germanium. *Physical Review*, **114**, 80–89.

127 Roth, L. M., Lax, B., and Zwerdling, S. (1959) Theory of Optical Magneto-Absorption Effects in Semiconductors. *Physical Review*, **114**, 90–104.

128 Luttinger, J. M. (1956) Quantum Theory of Cyclotron Resonance in Semiconductors: General Theory. *Physical Review*, **102**, 1030–1041.

129 Luttinger, J. M. and Kohn, W. (1955) Motion of Electrons and Holes in Perturbed Periodic Fields. *Physical Review*, **97**, 869–883.

130 Burt, M. G. (1992) The justification for applying the effective-mass approximation to microstructures. *Journal of Physics Condensed Matter*, **4**, 6651–6690.

131 Berezovsky, J., Ouyang, M., Meier, F., Awschalom, D. D., Battaglia, D., and Peng, X. (2005) Spin dynamics and level structure of quantum-dot quantum wells. *Physical Review B*, **71**, 081309.

132 Berezovsky, J., Gywat, O., Meier, F., Battaglia, D., Peng, X., and Awschalom, D. D. (2006) Initialization and read-out of spins in coupled core-shell quantum dots. *Nature Physics*, **2**, 831–834.

133 Xia, J.-B. (1989) Electronic structures of zero-dimensional quantum wells. *Physical Review B*, **40**, 8500–8507.

134 Meier, F. and Awschalom, D. D. (2005) Faraday rotation spectroscopy of quantum-dot quantum wells. *Physical Review B*, **71**, 205315.

135 Efros, Al. L. and Rosen, M. (1998) Quantum size level structure of narrow-gap semiconductor nanocrystals: Effect of band coupling. *Physical Review B*, **58**, 7120–7135.

136 Fock, V. (1928) Bemerkung zur Quantelung des harmonischen Oszillators im Magnetfeld. *Zeitschrift für Physik*, **47**, 446–448.

137 Darwin, C. G. (1931) The Diamagnetism of the Free Electron. *Mathematical Proceedings of the Cambridge Philosophical Society*, **27**, 86–90.

138 Burkard, G., Loss, D., and DiVincenzo, D. P. (1999) Coupled quantum dots as quantum gates. *Physical Review B*, **59**, 2070–2078.

139 Gywat, O., Burkard, G., and Loss, D. (2002) Biexcitons in coupled quantum dots as a source of entangled photons. *Physical Review B*, **65**, 205329.

140 Burkard, G., Seelig, G., and Loss, D. (2000) Spin interactions and switching in vertically tunnel-coupled quantum dots. *Physical Review B*, **62**, 2581–2592.

141 Borri, P., Langbein, W., Schneider, S., Woggon, U., Sellin, R. L., Ouyang, D., and Bimberg, D. (2001) Ultralong dephasing time in InGaAs quantum dots. *Physical Review Letters*, **87**, 157401.

142 Borri, P., Langbein, W., Schneider, S., Woggon, U., Sellin, R. L., Ouyang, D., and Bimberg, D. (2002) Relaxation and dephasing of multiexcitons in semiconductor quantum dots. *Physical Review Letters*, **89**, 187401.

143 Finley, J. J., Ashmore, A. D., Lemaitre, A., Mowbray, D. J., Skolnick, M. S., Itskevich, I. E., Maksym, P. A., Hopkinson, M., and Krauss, T. F. (2001) Charged and neutral exciton complexes in individual self-assembled In(Ga)As quantum dots. *Physical Review B*, **63**, 073307.

144 Regelman, D. V., Dekel, E., Gershoni, D., Ehrenfreund, E., Williamson, A. J., Shumway, J., Zunger, A., Schoenfeld, W. V., and Petroff, P. M. (2001) Optical spectroscopy of single quantum dots at tunable positive, neutral, and negative charge states. *Physical Review B*, **64**, 165301.

145 Rodt, S., Heitz, R., Schliwa, A., Sellin, R. L., Guffarth, F., and Bimberg, D. (2003) Repulsive exciton-exciton interaction in quantum dots. *Physical Review B*, **68**, 035331.

146 Rodt, S., Schliwa, A., Potschke, K., Guffarth, F., and Bimberg, D. (2005) Correlation of structural and few-particle properties of self-organized InAs/GaAs quantum dots. *Physical Review B*, **71**, 155325.

147 Drexler, H., Leonard, D., Hansen, W., Kotthaus, J. P., and Petroff, P. M. (1994) Spectroscopy of quantum levels in charge-tunable InGaAs quantum dots. *Physical Review Letters*, **73**, 2252–2255.

148 Bhattacharya, P., Ghosh, S., and Stiff-Roberts, A. D. (2004) Quantum Dot Opto-Electronic Devices. *Annual Review of Materials Research*, **34**, 1–40.

149 Benson, O., Santori, C., Pelton, M., and Yamamoto, Y. (2000) Regulated and entangled photons from a single quantum dot. *Physical Review Letters*, **84**, 2513–2516.

150 Shields, A. J. (2007) Semiconductor quantum light sources. *Nature Photonics*, **1**, 215–223.

151 Finley, J. J. *et al.* (2004) Quantum-confined stark shifts of charged exciton complexes in quantum dots. *Physical Review B*, **70**, 201308.

152 Reuter, D., Kailuweit, P., Wieck, A. D., Zeitler, U., Wibbelhoff, O., Meier, C., Lorke, A., and Maan, J. C. (2005) Coulomb-Interaction-Induced Incomplete Shell Filling in the Hole System of InAs Quantum Dots. *Physical Review Letters*, **94**, 026808.

153 He, L., Bester, G., and Zunger, A. (2005) Electronic Phase Diagrams of Carriers in Self-Assembled Quantum Dots: Violation of Hund's Rule and the Aufbau Principle for Holes. *Physical Review Letters*, **95**, 246804.

154 Bester, G., Reuter, D., He, L., Zunger, A., Kailuweit, P., Wieck, A. D., Zeitler, U., Maan, J. C., Wibbelhoff, O., and Lorke, A. (2007) Experimental imaging and atomistic modeling of electron and hole quasiparticle wave functions in InAs/GaAs quantum dots. *Physical Review B*, **76**, 075338.

155 Warburton, R. J., Durr, C. S., Karrai, K., Kotthaus, J. P., Medeiros-Ribeiro, G., and Petroff, P. M. (1997) Charged excitons in self-assembled semiconductor quantum dots. *Physical Review Letters*, **79**, 5282–5285.

156 Seidl, S. *et al.* (2005) Absorption and photoluminescence spectroscopy on a single self-assembled charge-tunable quantum dot. *Physical Review B*, **72**, 195339.

157 Findeis, F., Baier, M., Zrenner, A., Bichler, M., Abstreiter, G., Hohenester, U., and Molinari, E. (2001) Optical excitations of a self-assembled artificial ion. *Physical Review B*, **63**, 121309.

158 Baier, M., Findeis, F., Zrenner, A., Bichler, M., and Abstreiter, G. (2001) Optical spectroscopy of charged excitons in single quantum dot photodiodes. *Physical Review B*, **64**, 195326.

159 Ware, M. E., Stinaff, E. A., Gammon, D., Doty, M. F., Bracker, A. S., Gershoni, D., Korenev, V. L., Badescu, S. C., Lyanda-Geller, Y., and Reinecke, T. L. (2005) Polarized fine structure in the photoluminescence excitation spectrum of a negatively charged quantum dot. *Physical Review Letters*, **95**, 177403.

160 Finley, J. J. *et al.* (2001) Observation of multicharged excitons and biexcitons in a single InGaAs quantum dot. *Physical Review B*, **63**, 161305.

161 Atatüre, M., Dreiser, J., Badolato, A., Högele, A., Karrai, K., and Imamoğlu, A. (2006) Quantum-Dot Spin-State Preparation with Near-Unity Fidelity. *Science*, **312**, 551–553.

162 Laurent, S., Eble, B., Krebs, O., Lemaitre, A., Urbaszek, B., Marie, X., Amand, T., and Voisin, P. (2005) Electrical control of hole spin relaxation in charge tunable InAs/GaAs quantum dots. *Physical Review Letters*, **94**, 147401.

163 Dreiser, J., Atatüre, M., Galland, C., Müller, T., Badolato, A., and Imamoğlu, A. (2008) Optical investigations of quantum dot spin dynamics as a function of external electric and magnetic fields. *Physical Review B*, **77**, 075317.

164 Miller, D. A. B., Chemla, D. S., Damen, T. C., Gossard, A. C., Wiegmann, W., Wood, T. H., and Burrus, C. A. (1984) Band-Edge Electroabsorption in Quantum Well Structures: The Quantum-Confined Stark Effect. *Physical Review Letters*, **53**, 2173–2176.

165 Fry, P. W. *et al.* (2000) Inverted electron-hole alignment in InAs-GaAs self-assembled quantum dots. *Physical Review Letters*, **84**, 733–736.

166 Findeis, F., Baier, M., Beham, E., Zrenner, A., and Abstreiter, G. (2001) Photocurrent and photoluminescence of a single self-assembled quantum dot in electric fields. *Applied Physics Letters*, **78**, 2958–2960.

167 Alen, B., Bickel, F., Karrai, K., Warburton, R. J., and Petroff, P. M. (2003) Stark-shift modulation absorption spectroscopy of single quantum dots. *Applied Physics Letters*, **83**, 2235–2237.

168 Högele, A., Seidl, S., Kroner, M., Karrai, K., Warburton, R. J., Gerardot, B. D., and Petroff, P. M. (2004) Voltage-controlled optics of a quantum dot. *Physical Review Letters*, **93**, 217401.

169 Stufler, S., Ester, P., Zrenner, A., and Bichler, M. (2004) Power broadening of the exciton linewidth in a single In-GaAs/GaAs quantum dot. *Applied Physics Letters*, **85**, 4202–4204.

170 Polland, H.-J., Schultheis, L., Kuhl, J., Göbel, E. O., and Tu, C. W. (1985) Lifetime enhancement of two-dimensional excitons by the quantum-confined stark effect. *Physical Review Letters*, **55**, 2610–2613.

171 Bastard, G., Mendez, E. E., Chang, L. L., and Esaki, L. (1983) Variational calculations on a quantum well in an electric field. *Physical Review B*, **28**, 3241–3245.

172 Brum, J. A. and Bastard, G. (1985) Electric-field-induced dissociation of excitons in semiconductor quantum wells. *Physical Review B*, **31**, 3893–3898.

173 Mowbray, D. J. and Skolnick, M. S. (2005) New physics and devices based on self-assembled semiconductor quantum dots. *Journal of Physics D Applied Physics*, **38**, 2059–2076.

174 Born, M. and Wolf, E. (1959) *Principles of Optics: electromagnetic theory of propagation, interference, and diffraction of light*. Pergamon Press, New York.

175 Li, Y. Q., Steuerman, D. W., Berezovsky, J., Seferos, D. S., Bazan, G. C., and Awschalom, D. D. (2006) Cavity enhanced Faraday rotation of semiconductor quantum dots. *Applied Physics Letters*, **88**, 193126.

176 Ghosh, S., Wang, W., Mendoza, F., Myers, R., Li, X., Samarth, N., Gossard, A., and Awschalom, D. (2006) Enhancement of spin coherence using Q-factor engineering in semiconductor microdisc lasers. *Nature Materials*, **5**, 261–264.

177 Akahane, Y., Asano, T., Song, B., and Noda, S. (2003) High-Q photonic nanocavity in a two-dimensional photonic crystal. *Nature (London)*, **425**, 944 – 947.

178 Hennessy, K., Badolato, A., Winger, M., Gerace, D., Atatüre, M., Gulde, S., Fält, S., Hu, E. L., and Imamoğlu, A. (2007) Quantum nature of a strongly coupled single quantum dot-cavity system. *Nature (London)*, **445**, 896–899.

179 Imamoğlu, A., Fält, S., Dreiser, J., Fernandez, G., Atatüre, M., Hennessy, K., Badolato, A., and Gerace, D. (2007) Coupling quantum dot spins to a photonic crystal nanocavity. *Journal of Applied Physics*, **101**, 081602.

180 Clark, S. M., Fu, K.-M. C., Ladd, T. D., and Yamamoto, Y. (2007) Quantum computers based on electron spins controlled by ultrafast off-resonant single optical pulses. *Physical Review Letters*, **99**, 040501.

181 Imamoğlu, A., Awschalom, D. D., Burkard, G., DiVincenzo, D. P., Loss, D., Sherwin, M., and Small, A. (1999) Quantum information processing using quantum dot spins and cavity QED. *Physical Review Letters*, **83**, 4204–4207.

182 Mandel, L. and Wolf, E. (1995) *Optical Coherence and Quantum Optics*. Cambridge University Press, Cambridge.

183 Cohen-Tannoudji, C., Diu, B., and Laloë, F. (1977) *Quantum Mechanics, Volume 2*. John Wiley & Sons.

184 Jaynes, E. and Cummings, F. (1963) Comparison of quantum and semiclassical radiation theories with application to the beam maser. *Proceedings of the IEEE*, **51**, 89–109.

185 Raimond, J. M., Brune, M., and Haroche, S. (2001) Manipulating quantum entanglement with atoms and photons in a cavity. *Reviews of Modern Physics*, **73**, 565–582.

186 Carmichael, H. J. (1999) *Statistical Methods in Quantum Optics 1*. Springer, Berlin.

187 Gardiner, C. W. and Zoller, P. (2000) *Quantum Noise*. Springer, Berlin.

188 Gywat, O., Engel, H.-A., Loss, D., Epstein, R. J., Mendoza, F. M., and Awschalom, D. D. (2004) Optical detection of single-electron spin decoherence in a quantum dot. *Physical Review B*, **69**, 205303.

189 Gywat, O., Engel, H.-A., and Loss, D. (2005) Probing single-electron spin decoherence in quantum dots using charged excitons. *Journal of Superconductivity and Novel Magnetism*, **18**, 175.

190 Golovach, V. N., Khaetskii, A., and Loss, D. (2004) Phonon-induced decay of the electron spin in quantum dots. *Physical Review Letters*, **93**, 016601.

191 Reithmaier, J. P., Sek, G., Loffler, A., Hofmann, C., Kuhn, S., Reitzenstein, S., Keldysh, L. V., Kulakovskii, V. D., Reinecke, T. L., and Forchel, A. (2004) Strong coupling in a single quantum dot-semiconductor microcavity system. *Nature (London)*, **432**, 197–200.

192 Yoshie, T., Scherer, A., Hendrickson, J., Khitrova, G., Gibbs, H. M., Rupper, G., Ell, C., Shchekin, O. B., and Deppe, D. G. (2004) Vacuum Rabi splitting with a single quantum dot in a photonic crystal nanocavity. *Nature (London)*, **432**, 200–203.

193 Englund, D., Faraon, A., Fushman, I., Stoltz, N., Petroff, P., and Vuckovic, J. (2007) Controlling cavity reflectivity with a single quantum dot. *Nature (London)*, **450**, 857–861.

194 Srinivasan, K. and Painter, O. (2007) Linear and nonlinear optical spectroscopy of a strongly coupled microdisk-quantum dot system. *Nature (London)*, **450**, 862–865.

195 van Enk, S. J., Cirac, J. I., and Zoller, P. (1997) Ideal quantum communication over noisy channels: a quantum optical implementation. *Physical Review Letters*, **78**, 4293–4296.

196 Imamoğlu, A. (2000) Quantum computation using quantum dot spins and microcavities. *Fortschritte der Physik*, **48**, 987–997.

197 Blais, A., Huang, R.-S., Wallraff, A., Girvin, S. M., and Scheolkopf, R. J. (2004) Cavity quantum electrodynamics for superconducting electrical circuits: An architecture for quantum computation. *Physical Review A*, **69**, 062320.

198 Purcell, E. M. (1946) Spontaneous emission probabilities at radio frequencies. *Physical Review*, **69**, 681.

199 Kiraz, A., Michler, P., Becher, C., Gayral, B., Imamoğlu, A., Zhang, L., Hu, E., Schoenfeld, W. V., and Petroff, P. M. (2001) Cavity-quantum electrodynamics using a single InAs quantum dot in a microdisk structure. *Applied Physics Letters*, **78**, 3932–3934.

200 Moreau, E., Robert, I., Gérard, J. M., Abram, I., Manin, L., and Thierry-Mieg, V. (2001) Single-mode solid-state single photon source based on isolated quantum dots in pillar microcavities. *Applied Physics Letters*, **79**, 2865–2867.

201 Solomon, G. S., Pelton, M., and Yamamoto, Y. (2001) Single-mode spontaneous emission from a single quantum dot in a three-dimensional microcavity. *Physical Review Letters*, **86**, 3903–3906.

202 Vuckovic, J., Fattal, D., Santori, C., Solomon, G. S., and Yamamoto, Y. (2003) Enhanced single-photon emission from a quantum dot in a micropost microcavity. *Applied Physics Letters*, **82**, 3596–3598.

203 Bayer, M., Reinecke, T. L., Weidner, F., Larionov, A., McDonald, A., and Forchel, A. (2001) Inhibition and enhancement of the spontaneous emission of quantum dots in structured microresonators. *Physical Review Letters*, **86**, 3168–3171.

204 Wallraff, A., Schuster, D. I., Blais, A., Frunzio, L., Huang, R.-S., Majer, J., Kumar, S., Girvin, S. M., and Schoelkopf, R. J. (2004) Strong coupling of a single photon to a superconducting qubit using circuit quantum electrodynamics. *Nature (London)*, **431**, 162–167.

205 Gywat, O., Meier, F., Loss, D., and Awschalom, D. D. (2006) Dynamics of coupled qubits interacting with an off-resonant cavity. *Physical Review B*, **73**, 125336.

206 van Kesteren, H. W., Cosman, E. C., van der Poel, W. A. J. A., and Foxon, C. T. (1990) Fine structure of excitons in type-II GaAs/AlAs quantum wells. *Physical Review B*, **41**, 5283–5292.

207 Bayer, M. *et al.* (2002) Fine structure of neutral and charged excitons in self-assembled In(Ga)As/(Al)GaAs quantum dots. *Physical Review B*, **65**, 195315.

208 Finley, J. J., Mowbray, D. J., Skolnick, M. S., Ashmore, A. D., Baker, C., Monte, A. F. G., and Hopkinson, M. (2002) Fine structure of charged and neutral excitons in InAs-$Al_{0.6}Ga_{0.4}As$ quantum dots. *Physical Review B*, **66**, 153316.

209 Kulakovskii, V. D., Bacher, G., Weigand, R., Kümmell, T., Forchel, A., Borovitskaya, E., Leonardi, K., and Hommel, D. (1999) Fine structure of biexciton emission in symmetric and asymmetric CdSe/ZnSe single quantum dots. *Physical Review Letters*, **82**, 1780–1783.

210 Takagahara, T. (2000) Theory of exciton doublet structures and polarization relaxation in single quantum dots. *Physical Review B*, **62**, 16840–16855.

211 Bracker, A. S. *et al.* (2005) Optical pumping of the electronic and nuclear spin of single charge-tunable quantum dots. *Physical Review Letters*, **94**, 047402.

212 Moreau, E., Robert, I., Manin, L., Thierry-Mieg, V., Gérard, J. M., and Abram, I. (2001) Quantum cascade of photons in semiconductor quantum dots. *Physical Review Letters*, **87**, 183601.

213 Aspect, A., Grangier, P., and Roger, G. (1981) Experimental Tests of Realistic Local Theories via Bell's Theorem. *Physical Review Letters*, **47**, 460–463.

214 Aspect, A., Grangier, P., and Roger, G. (1982) Experimental realization of Einstein–Podolsky–Rosen–Bohm Gedanken-experiment: a new violation of Bell's inequalities. *Physical Review Letters*, **49**, 91–94.

215 Aspect, A., Dalibard, J., and Roger, G. (1982) Experimental test of Bell's inequalities using time-varying analyzers. *Physical Review Letters*, **49**, 1804–1807.

216 Kiraz, A., Fälth, S., Becher, C., Gayral, B., Schoenfeld, W. V., Petroff, P. M., Zhang, L., Hu, E., and Imamoğlu, A. (2002) Photon correlation spectroscopy of a single quantum dot. *Physical Review B*, **65**, 161303.

217 Santori, C., Fattal, D., Pelton, M., Solomon, G. S., and Yamamoto, Y. (2002) Polarization-correlated photon pairs from a single quantum dot. *Physical Review B*, **66**, 045308.

218 Stevenson, R. M., Thompson, R. M., Shields, A. J., Farrer, I., Kardynal, B. E., Ritchie, D. A., and Pepper, M. (2002) Quantum dots as a photon source for passive quantum key encoding. *Physical Review B*, **66**, 081302.

219 Zwiller, V., Jonsson, P., Blom, H., Jeppesen, S., Pistol, M.-E., Samuelson, L., Katznelson, A. A., Kotelnikov, E. Y., Evtikhiev, V., and Björk, G. (2002) Correlation spectroscopy of excitons and biexcitons on a single quantum dot. *Physical Review A*, **66**, 053814.

220 Ulrich, S. M., Strauf, S., Michler, P., Bacher, G., and Forchel, A. (2003) Triggered polarization-correlated photon pairs from a single CdSe quantum dot. *Applied Physics Letters*, **83**, 1848–1850.

221 Akopian, N., Lindner, N. H., Poem, E., Berlatzky, Y., Avron, J., Gershoni, D., Gerardot, B. D., and Petroff, P. M. (2006) Entangled Photon Pairs from Semiconductor Quantum Dots. *Physical Review Letters*, **96**, 130501.

222 Young, R. J., Stevenson, R. M., Atkinson, P., Cooper, K., Ritchie, D. A., and Shields, A. J. (2006) Improved fidelity of triggered entangled photons from single quantum dots. *New Journal of Physics*, **8**, 29.

223 Cerletti, V., Gywat, O., and Loss, D. (2005) Entanglement transfer from electron spins to photons in spin light-emitting diodes containing quantum dots. *Physical Review B*, **72**, 115316.

224 Flindt, C., Sørensen, A. S., Lukin, M. D., and Taylor, J. M. (2007) Spin-photon entangling diode. *Physical Review Letters*, **98**, 240501.

225 Abragam, A. (1961) *Principles of Nuclear Magnetism*. Oxford University Press, Oxford.

226 Merkulov, I. A., Efros, Al. L., and Rosen, M. (2002) Electron spin relaxation by nuclei in semiconductor quantum dots. *Physical Review B*, **65**, 205309.

227 Khaetskii, A. V., Loss, D., and Glazman, L. (2002) Electron spin decoherence in quantum dots due to interaction with nuclei. *Physical Review Letters*, **88**, 186802.

228 Coish, W. A. and Loss, D. (2004) Hyperfine interaction in a quantum dot: non-Markovian electron spin dynamics. *Physical Review B*, **70**, 195340.

229 Fischer, J., Coish, W. A., Bulaev, D. V., and Loss, D. (2008) Spin decoherence of a heavy hole coupled to nuclear spins in a quantum dot. *Physical Review B*, **78**, 155329.

230 Cortez, S., Krebs, O., Laurent, S., Senes, M., Marie, X., Voisin, P., Ferreira, R., Bastard, G., Gerard, J.-M., and Amand, T. (2002) Optically driven spin memory in n-doped InAs-GaAs quantum dots. *Physical Review Letters*, **89**, 207401.

231 Gammon, D., Efros, Al. L., Kennedy, T. A., Rosen, M., Katzer, D. S., Park, D., Brown, S. W., Korenev, V. L., and Merkulov, I. A. (2001) Electron and nuclear spin interactions in the optical spectra of single GaAs quantum dots. *Physical Review Letters*, **86**, 5176–5179.

232 Kroutvar, M., Ducommun, Y., Heiss, D., Bichler, M., Schuh, D., Abstreiter, G., and Finley, J. J. (2004) Optically programmable electron spin memory using semiconductor quantum dots. *Nature (London)*, **432**, 81–84.

233 Besombes, L., Leger, Y., Maingault, L., Ferrand, D., Mariette, H., and Cibert, J. (2004) Probing the spin state of a single magnetic ion in an individual quantum dot. *Physical Review Letters*, **93**, 207403.

234 Högele, A. *et al.* (2005) Spin-selective optical absorption of singly charged excitons in a quantum dot. *Applied Physics Letters*, **86**, 221905.

235 Epstein, R. J., Fuchs, D. T., Schoenfeld, W. V., Petroff, P. M., and Awschalom, D. D. (2001) Hanle effect measurements of spin lifetimes in InAs self-assembled quantum dots. *Applied Physics Letters*, **78**, 733–735.

236 Gupta, J. A., Awschalom, D. D., Peng, X., and Alivisatos, A. P. (1999) Spin coherence in semiconductor quantum dots. *Physical Review B*, **59**, R10421–R10424.

237 Greilich, A., Yakovlev, D. R., Shabaev, A., Efros, Al. L., Yugova, I. A., Oulton, R., Stavarache, V., Reuter, D., Wieck, A., and Bayer, M. (2006) Mode locking of electron spin coherences in singly charged quantum dots. *Science*, **313**, 341–345.

238 Mikkelsen, M. H., Berezovsky, J., Stoltz, N. G., Coldren, L. A., and Awschalom, D. D. (2007) Optically detected coherent spin dynamics of a single electron in a quantum dot. *Nature Physics*, **3**, 770–773.

239 Berezovsky, J., Mikkelsen, M. H., Stoltz, N. G., Coldren, L. A., and Awschalom, D. D. (2008) Picosecond coherent optical manipulation of a single electron spin in a quantum dot. *Science*, **320**, 349–352.

240 Press, D., Ladd, T. D., Zhang, B., and Yamamoto, Y. (2008) Complete quantum control of a single quantum dot spin using ultrafast optical pulses. *Nature (London)*, **456**, 218–221.

241 Kikkawa, J. M. and Awschalom, D. D. (1998) Resonant spin amplification in n-type GaAs. *Physical Review Letters*, **80**, 4313.

242 Kim, D., Economou, S. E., Ştefan C. Bădescu, Scheibner, M., Bracker, A. S., Bashkansky, M., Reinecke, T. L., and Gammon, D. (2008) Optical spin initialization and nondestructive measurement in a quantum dot molecule. *Physical Review Letters*, **101**, 236804.

243 Xu, X., Sun, B., Berman, P. R., Steel, D. G., Bracker, A. S., Gammon, D., and Sham, L. J. (2008) Coherent population trapping of an electron spin in a single negatively charged quantum dot. *Nature Physics*, **4**, 692.

244 Gerardot, B. D., Brunner, D., Dalgarno, P. A., Ohberg, P., Seidl, S., Kroner, M., Karrai, K., Stoltz, N. G., Petroff, P. M., and Warburton, R. J. (2008) Optical pumping of a single hole spin in a quantum dot. *Nature (London)*, **451**, 441–444.

245 Greilich, A., Shabaev, A., Yakovlev, D. R., Efros, Al. L., Yugova, I. A., Reuter, D., Wieck, A. D., and Bayer, M. (2007) Nuclei-Induced Frequency Focusing of Electron Spin Coherence. *Science*, **317**, 1896–1899.

246 Morton, J. J. L., Tyryshkin, A. M., Brown, R. M., Shankar, S., Lovett, B. W., Ardavan, A., Schenkel, T., Haller, E. E., Ager, J. W., and Lyon, S. A. (2008) Solid-state quantum memory using the ^{31}P nuclear spin. *Nature (London)*, **455**, 1085–1088.

247 Jelezko, F., Gaebel, T., Popa, I., Domhan, M., Gruber, A., and Wrachtrup, J. (2004) Observation of coherent oscillation of a single nuclear spin and realization of a two-qubit conditional quantum gate. *Physical Review Letters*, **93**, 130501.

248 Smyth, J. F., Tulchinsky, D. A., Awschalom, D. D., Samarth, N., Luo, H., and Furdyna, J. K. (1993) Femtosecond scattering dynamics in magnetic semiconductor spin superlattices. *Physical Review Letters*, **71**, 601–604.

249 Khaetskii, A. V. and Nazarov, Y. V. (2001) Spin-flip transitions between Zeeman sublevels in semiconductor quantum dots. *Physical Review B*, **64**, 125316.

250 Woods, L. M., Reinecke, T. L., and Lyanda-Geller, Y. (2002) Spin relaxation in quantum dots. *Physical Review B*, **66**, 161318.

251 Heiss, D., Schaeck, S., Huebl, H., Bichler, M., Abstreiter, G., Finley,

J. J., Bulaev, D. V., and Loss, D. (2007) Observation of extremely slow hole spin relaxation in self-assembled quantum dots. *Physical Review B*, **76**, 241306.

252 Heiss, D., Jovanov, V., Bichler, M., Abstreiter, G., and Finley, J. J. (2008) Charge and spin readout scheme for single self-assembled quantum dots. *Physical Review B*, **77**, 235442.

253 Ramsay, A. J., Boyle, S. J., Kolodka, R. S., Oliveira, J. B. B., Skiba-Szymanska, J., Liu, H. Y., Hopkinson, M., Fox, A. M., and Skolnick, M. S. (2008) Fast optical preparation, control, and readout of a single quantum dot spin. *Physical Review Letters*, **100**, 197401.

254 Myers, R. C., Mikkelsen, M. H., Tang, J. M., Gossard, A. C., Flatte, M. E., and Awschalom, D. D. (2008) Zero-field optical manipulation of magnetic ions in semiconductors. *Nature Materials*, **7**, 203–208.

255 Kudelski, A., Lemaitre, A., Miard, A., Voisin, P., Graham, T. C. M., Warburton, R. J., and Krebs, O. (2007) Optically probing the fine structure of a single Mn atom in an InAs quantum dot. *Physical Review Letters*, **99**, 247209.

256 Vamivakas, A. N., Atatüre, M., Dreiser, J., Yilmaz, S. T., Badolato, A., Swan, A. K., Goldberg, B. B., Imamoğlu, A., and Uenlue, M. S. (2007) Strong extinction of a far-field laser beam by a single quantum dot. *Nano Letters*, **7**, 2892–2896.

257 Kroner, M., Weiss, K. M., Biedermann, B., Seidl, S., Holleitner, A. W., Badolato, A., Petroff, P. M., Öhberg, P., Warburton, R. J., and Karrai, K. (2008) Resonant two-color high-resolution spectroscopy of a negatively charged exciton in a self-assembled quantum dot. *Physical Review B*, **78**, 075429.

258 Dzhioev, R. I., Korenev, V. L., Zakharchenya, B. P., Gammon, D., Bracker, A. S., Tischler, J. G., and Katzer, D. S. (2002) Optical orientation and the Hanle effect of neutral and negatively charged excitons in GaAs/$Al_x Ga_{1-x}$As quantum wells. *Physical Review B*, **66**, 153409.

259 Gupta, J. A., Awschalom, D. D., Efros, Al. L., and Rodina, A. V. (2002) Spin dynamics in semiconductor nanocrystals. *Physical Review B*, **66**, 125307.

260 Koppens, F. H. L., Buizert, C., Tielrooij, K. J., Vink, I. T., Nowack, K. C., Meunier, T., Kouwenhoven, L. P., and Vandersypen, L. M. K. (2006) Driven coherent oscillations of a single electron spin in a quantum dot. *Nature (London)*, **442**, 766–771.

261 Kroner, M. *et al.* (2008) Optical detection of single-electron spin resonance in a quantum dot. *Physical Review Letters*, **100**, 156803.

262 Salis, G. and Moser, M. (2005) Faraday-rotation spectrum of electron spins in microcavity-embedded GaAs quantum wells. *Physical Review B*, **72**, 115325.

263 Kupriyanov, D. V. and Sokolov, I. M. (1992) Optical-detection of magnetic-resonance by classical and squeezed light. *Quantum Optics*, **4**, 55–70.

264 Guest, J. R. *et al.* (2002) Measurement of optical absorption by a single quantum dot exciton. *Physical Review B*, **65**, 241310.

265 Awschalom, D. D., Loss, D., and Samarth, N. (eds) (2002) *Semiconductor Spintronics and Quantum Computation*. NanoScience and Technology, Springer-Verlag.

266 Gurudev-Dutt, M. V. *et al.* (2005) Stimulated and spontaneous optical generation of electron spin coherence in charged GaAs quantum dots. *Physical Review Letters*, **94**, 227403.

267 Semenov, Y. G. and Kim, K. W. (2004) Phonon-mediated electron-spin phase diffusion in a quantum dot. *Physical Review Letters*, **92**, 026601.

268 Erlingsson, S. I., Nazarov, Y. V., and Fal'ko, V. I. (2001) Nucleus-mediated spin-flip transitions in GaAs quantum dots. *Physical Review B*, **64**, 195306.

269 Meier, F. and Awschalom, D. D. (2004) Spin-photon dynamics of quantum dots in two-mode cavities. *Physical Review B*, **70**, 205329.

270 Leuenberger, M. N. (2006) Fault-tolerant quantum computing with coded spins using the conditional Faraday rotation in quantum dots. *Physical Review B*, **73**, 075312.

271 Combescot, M. and Betbeder-Matibet, O. (2004) Theory of spin precession moni-

tored by laser pulse. *Solid State Communications*, **132**, 129–134.

272 Chen, P., Piermarocchi, C., Sham, L. J., Gammon, D., and Steel, D. G. (2004) Theory of quantum optical control of a single spin in a quantum dot. *Physical Review B*, **69**, 075320.

273 Pryor, C. E. and Flatté, M. E. (2006) Predicted ultrafast single-qubit operations in semiconductor quantum dots. *Applied Physics Letters*, **88**, 233108.

274 Economou, S. E., Sham, L. J., Wu, Y., and Steel, D. G. (2006) Proposal for optical U(1) rotations of electron spin trapped in a quantum dot. *Physical Review B*, **74**, 205415.

275 Cohen-Tannoudji, C. and Dupont-Roc, J. (1972) Experimental study of Zeeman light shifts in weak magnetic fields. *Physical Review A*, **5**, 968–984.

276 Cohen-Tannoudji, C. and Reynaud, S. (1977) Dressed-atom description of resonance fluorescence and absorption spectra of a multi-level atom in an intense laser beam. *Journal of Physics B*, **10**, 345–363.

277 Suter, D., Klepel, H., and Mlynek, J. (1991) Time-resolved two-dimensional spectroscopy of optically driven atomic sublevel coherences. *Physical Review Letters*, **67**, 2001–2004.

278 Combescot, M. and Combescot, R. (1988) Excitonic Stark shift: A coupling to "semivirtual" biexcitons. *Physical Review Letters*, **61**, 117–120.

279 Joffre, M., Hulin, D., Migus, A., and Combescot, M. (1989) Laser-induced exciton splitting. *Physical Review Letters*, **62**, 74–77.

280 Papageorgiou, G., Chari, R., Brown, G., Kar, A. K., Bradford, C., Prior, K. A., Kalt, H., and Galbraith, I. (2004) Spectral dependence of the optical Stark effect in ZnSe-based quantum wells. *Physical Review B*, **69**, 085311.

281 Gupta, J. A., Knobel, R., Samarth, N., and Awschalom, D. D. (2001) Ultrafast manipulation of electron spin coherence. *Science*, **292**, 2458–2461.

282 Unold, T., Mueller, K., Lienau, C., Elsaesser, T., and Wieck, A. D. (2005) Optical control of excitons in a pair of quantum dots coupled by the dipole-dipole

interaction. *Physical Review Letters*, **94**, 137404.

283 Dutt, M. V. G., Cheng, J., Wu, Y., Xu, X., Steel, D. G., Bracker, A. S., Gammon, D., Economou, S. E., Liu, R.-B., and Sham, L. J. (2006) Ultrafast optical control of electron spin coherence in charged GaAs quantum dots. *Physical Review B*, **74**, 125306.

284 Wu, Y., Kim, E. D., Xu, X., Cheng, J., Steel, D. G., Bracker, A. S., Gammon, D., Economou, S. E., and Sham, L. J. (2007) Selective optical control of electron spin coherence in singly charged GaAs-Al$_{0.3}$Ga$_{0.7}$As quantum dots. *Physical Review Letters*, **99**, 097402.

285 Klingshirn, C. (2006) *Semiconductor Optics*. Springer, 3rd edn.

286 von Lehmen, A., Zucker, J. E., Heritage, J. P., and Chemla, D. S. (1987) Phonon sideband of quasi-two-dimensional excitons in GaAs quantum wells. *Physical Review B*, **35**, R6479.

287 Rosatzin, M., Suter, D., and Mlynek, J. (1990) Light-shift-induced spin echoes in a J = 1/2 atomic ground state. *Physical Review A*, **42**, R1839–R1841.

288 Ouyang, M. and Awschalom, D. D. (2003) Coherent Spin Transfer Between Molecularly Bridged Quantum Dots. *Science*, **301**, 1074–1078.

289 Meier, F., Cerletti, V., Gywat, O., Loss, D., and Awschalom, D. D. (2004) Molecular spintronics: Coherent spin transfer in coupled quantum dots. *Physical Review B*, **69**, 195315.

290 Schrier, J. and Whaley, K. B. (2005) Atomistic theory of coherent spin transfer between molecularly bridged quantum dots. *Physical Review B*, **72**, 085320.

291 Govorov, A. O. (2003) Spin and energy transfer in nanocrystals without tunneling. *Physical Review B*, **68**, 075315.

292 Nazir, A., Lovett, B. W., Barrett, S. D., Reina, J. H., and Briggs, G. A. D. (2005) Anticrossings in Förster coupled quantum dots. *Physical Review B*, **71**, 045334.

293 Stinaff, E. A., Scheibner, M., Bracker, A. S., Ponomarev, I. V., Korenev, V. L., Ware, M. E., Doty, M. F., Reinecke, T. L., and Gammon, D. (2006) Optical signatures of coupled quantum dots. *Science*, **311**, 636–639.

294 Doty, M. F., Scheibner, M., Ponomarev, I. V., Stinaff, E. A., Bracker, A. S., Korenev, V. L., Reinecke, T. L., and Gammon, D. (2006) Electrically tunable g factors in quantum dot molecular spin states. *Physical Review Letters*, **97**, 197202.

295 Scheibner, M., Doty, M. F., Ponomarev, I. V., Bracker, A. S., Stinaff, E. A., Korenev, V. L., Reinecke, T. L., and Gammon, D. (2007) Spin fine structure of optically excited quantum dot molecules. *Physical Review B*, **75**, 245318.

296 Scheibner, M., Ponomarev, I. V., Stinaff, E. A., Doty, M. F., Bracker, A. S., Hellberg, C. S., Reinecke, T. L., and Gammon, D. (2007) Photoluminescence spectroscopy of the molecular biexciton in vertically stacked InAs-GaAs quantum dot pairs. *Physical Review Letters*, **99**, 197402.

297 Heitler, W. and London, F. (1927) Wechselwirkung neutraler Atome und homöopolare Bindung nach der Quantenmechanik. *Zeitschrift für Physik*, **44**, 455–472.

298 Krenner, H. J., Clark, E. C., Nakaoka, T., Bichler, M., Scheurer, C., Abstreiter, G., and Finley, J. J. (2006) Optically probing spin and charge interactions in a tunable artificial molecule. *Physical Review Letters*, **97**, 076403.

299 Doty, M. F., Climente, J. I., Korkusinski, M., Scheibner, M., Bracker, A. S., Hawrylak, P., and Gammon, D. (2009) Antibonding ground states in InAs quantum-dot molecules. *Physical Review Letters*, **102**, 047401.

300 Türeci, H. E., Taylor, J. M., and Imamoğlu, A. (2007) Coherent optical manipulation of triplet-singlet states in coupled quantum dots. *Physical Review B*, **75**, 235313.

301 Economou, S. E. and Reinecke, T. L. (2008) Optically induced spin gates in coupled quantum dots using the electron–hole exchange interaction. *Physical Review B*, **78**, 115306.

302 Doty, M. F., Scheibner, M., Bracker, A. S., Ponomarev, I. V., Reinecke, T. L., and Gammon, D. (2008) Optical spectra of doubly charged quantum dot molecules in electric and magnetic fields. *Physical Review B*, **78**, 115316.

303 Fält, S., Atatüre, M., Türeci, H. E., Zhao, Y., Badolato, A., and Imamoğlu, A. (2008) Strong electron–hole exchange in coherently coupled quantum dots. *Physical Review Letters*, **100**, 106401.

304 Robledo, L., Elzerman, J., Jundt, G., Atature, M., Hogele, A., Falt, S., and Imamoğlu, A. (2008) Conditional dynamics of interacting quantum dots. *Science*, **320**, 772–775.

305 Andlauer, T. and Vogl, P. (2009) Electrically controllable g tensors in quantum dot molecules. *Physical Review B*, **79**, 045307.

306 Badolato, A., Hennessy, K., Atatüre, M., Dreiser, J., Hu, E., Petroff, P. M., and Imamoğlu, A. (2005) Deterministic coupling of single quantum dots to single nanocavity modes. *Science*, **308**, 1158–1161.

307 Laucht, A., Hofbauer, F., Hauke, N., Angele, J., Stobbe, S., Kaniber, M., Bohm, G., Lodahl, P., Amann, M. C., and Finley, J. J. (2009) Electrical control of spontaneous emission and strong coupling for a single quantum dot. *New Journal of Physics*, **11**, 023034.

308 Mohr, P. J., Taylor, B. N., and Newell, D. B. (2008) CODATA recommended values of the fundamental physical constants: 2006. *Reviews of Modern Physics*, **80**, 633.

309 Vurgaftman, I., Meyer, J. R., and Ram-Mohan, L. R. (2001) Band parameters for III–V compound semiconductors and their alloys. *Journal of Applied Physics*, **89**, 5815–5875.

310 Willatzen, M., Cardona, M., and Christensen, N. E. (1995) Spin-orbit coupling parameters and electron g factor of II–VI zinc-blende materials. *Physical Review B*, **51**, 17992–17994.

311 Shen, K., Weng, M. Q., and Wu, M. W. (2008) L-valley electron g-factor in bulk GaAs and AlAs. *Journal of Applied Physics*, **104**, 063719.

312 Karimov, O. Z., Wolverson, D., Davies, J. J., Stepanov, S. I., Ruf, T., Ivanov, S. V., Sorokin, S. V., O'Donnell, C. B., and Prior, K. A. (2000) Electron g-factor for cubic $Zn_{1-x}Cd_x$Se determined by spin-flip Raman scattering. *Physical Review B*, **62**, 16582–16586.

313 Weisbuch, C. and Hermann, C. (1977) Optical detection of conduction-electron spin resonance in GaAs, $Ga_{1-x}In_xAs$, and $Ga_{1-x}Al_xAs$. *Physical Review B*, **15**, 816–822.

314 Hermann, C. and Weisbuch, C. (1977) $\vec{k}\cdot\vec{p}$ perturbation theory in III–V compounds and alloys: a reexamination. *Physical Review B*, **15**, 823–833.

Index

Spins in Optically Active Quantum Dots. Concepts and Methods.
Oliver Gywat, Hubert J. Krenner, and Jesse Berezovsky
Copyright © 2010 WILEY-VCH Verlag GmbH & Co. KGaA, Weinheim
ISBN: 978-3-527-40806-1